W9-ACS-046

CHROMATOGRAPHY IN GEOLOGY

SERIES

Methods in Geochemistry and Geophysics

CHROMATOGRAPHY IN GEOLOGY

BY

ARTHUR S. RITCHIE
Department of Geology, Newcastle University College, Australia

ELSEVIER PUBLISHING COMPANY
AMSTERDAM / LONDON / NEW YORK
1964

ELSEVIER PUBLISHING COMPANY
335 JAN VAN GALENSTRAAT, P.O. BOX 211, AMSTERDAM

AMERICAN ELSEVIER PUBLISHING COMPANY, INC.
52 VANDERBILT AVENUE, NEW YORK 17, N.Y.

ELSEVIER PUBLISHING COMPANY LIMITED
12B, RIPPLESIDE COMMERCIAL ESTATE
RIPPLE ROAD, BARKING, ESSEX

LIBRARY OF CONGRESS CATALOG CARD NUMBER 64-11341

WITH 5 ILLUSTRATIONS, 2 PLATES AND 41 TABLES

Preface

When the techniques of one science are applied to a related field, there comes a time when the applications, scattered through the scientific literature, need to be collected and unified into one volume. When the publishers suggested to me that the time had arrived for such a collection of applications of chromatography to geology and that I should make the collection, I agreed to each suggestion — to the first objectively and to the second with personally biassed subjectivity.

One man rarely writes a book unaided. In the preparation of this manuscript I have been fortunate in having received assistance in so many ways. The Government of France provided not only financial assistance but also the privilege of my working at the Radium Institute, Paris in 1959–1960 under the guidance and with the inspiration of Dr. M. Lederer. The Journal of Chemical Education and the Economic Geology Publishing Company kindly permitted the author to include in this book data formerly published by him in their journals. The United States Geological Survey and authors such as H. Agrinier, R. Coulomb, H. Nevill and R. Lever, A. Snelgrove and his colleagues and A. A. North and his colleagues gave the author permission to quote from and paraphrase their applications of chromatography to geology. Constructive criticism and discussion of the manuscript have been given by my colleague Dr. W. Pickering. M. Stemprok of Prague and F. Hecht of Vienna assisted by supplying reprints from their countries. Stenographic assistance has been rendered by Miss M. Hews, Miss B. Cook and Mrs. D. Cahill. The draughting of the diagrams was the work of Miss D. Frost. Proof-reading and assistance in many ways were rendered by my wife and children. My university and my colleagues aided in providing facilities, encouragement and advice. Guidance and encouragement were given readily by the publishers. The author gratefully acknowledges all the assistance which he has received and without which this task could not have been completed.

Newcastle, Australia
February 1963

A. S. RITCHIE

Contents

CHAPTER 1

Introduction

The geologist, and more especially the geochemist, must combine his geological knowledge in the collection and preparation of rocks and minerals for analysis, and in the interpretation of results, with either his or another's skill in chemical and physical analytical techniques. When a great new field of chemical analysis is developed, disinterest by the geologist cannot be justified: applications of the new technique must and will be found.

The impact of the chromatographic technique on both organic and inorganic chemical studies has been spectacular (LEDERER and LEDERER, 1957; MILTON and WATERS, 1955). Several whole fields of specialization for chemists have emerged in the last decade. An important characteristic of the method is its applicability to fields of science related to chemistry. In a *limited way*, related scientists can become their own chemists.

It is not surprising that the rapid development of chromatography in the fields of chemistry has left some geochemists and most geologists wanting a starting point and some guidance on a path to lead them to fruitful applications of the method to geology. Having found a starting point and having progressed a little way, the author hopes that this book will guide others along and beyond his path.

The book has been written firstly for those geologists who have been saying "I should do something about applying chromatography to my problem". Secondly, it is hoped that, by its publication, the book will show others the possibility of using the method. In a third way, it is hoped that the book will provide a summary of many of the applications of chromatography to geology which, at present, are scattered through the literature.

The applications of chromatography to geology are already

numerous. Some methods fall distinctly within the field of chemistry, others likewise in geology, while some lie on the border between the two subjects. Since in the field of science there are no boundaries, the author has considered here only those applications which are normally carried out by geologists or geochemists or which could be carried out by them. Thus a laboratory chromatographic method involving the separation of the rare earths (M. LEDERER, 1953) has not been included, whereas a laboratory method for the estimation of molybdenum in ore has (AGRINIER, 1959).

The working skills in chromatography are not difficult to learn. Care has been taken here to present the techniques separately from the theory. It is possible to analyse by the chromatographic method without understanding the processes that achieve the results. This is true also of other analytical methods but, nevertheless, it would be foolish to deny that better results, and certainly more ingenious applications, will come from those who understand more fully what they are doing. For this reason, a chapter on the theory of the chromatographic processes has been included. The author has attempted to present the theory in a more simplified way than that found in publications written for chemists. It is hoped that accuracy has not been sacrificed unduly by so doing.

Chromatography is regarded by chemists mostly as an analytical method. Thus from LEDERER and LEDERER (1957) we read "chromatography is an analytical method for the purification and separation of organic and inorganic substances". GORDON et al. (1943) defined it as "the technical procedure of analysis by percolation of a fluid through a body of comminuted or porous rigid material, irrespective of the nature of the physico-chemical processes that may lead to the separation of substances in the apparatus". CASSIDY (1951, p.208) appears to be more restrictive by defining it thus: "Chromatography is essentially a method for the countercurrent application of adsorption". A less restrictive approach might be that chromatography embraces the physico-chemical processes of partition, adsorption, diffusion and ion-exchange.

In its original sense and use by Tswett (as quoted by ZECHMEISTER and CHOLNOKY, 1950, p.1), chromatography was a method of separation of coloured and colourless substances achieved by firstly placing them on adsorbent columns and then removing them selectively by chosen solvents. The introduction of *paper chromatography* in 1943 (GORDON et al., 1943) permitted a greater variety of separations from extremely minute samples. More recently, other thin-layer materials have been introduced to add further variety and adaptability to chromatography.

It seems appropriate that further definitive discussion be left until some considerations of the theory of the chromatographic processes have been made. The field of gas chromatography is not considered in this book, although recent work indicates that there are interesting geological applications, e.g., in the analysis of the rare gases, in the analysis of volatile hydrocarbons, etc., etc. For these we refer the reader to the numerous books dealing with gas chromatography (SCHAY, 1960; AMBROSE and AMBROSE, 1961; BAYER, 1961; DAL NOGARE and JUVET, 1962; LITTLEWOOD, 1962).

CHAPTER 2

The Chromatographic Processes

The phenomena of partition, diffusion, adsorption and ion-exchange have been recognised, defined and studied initially without reference to chromatography. Nevertheless, since these phenomena (collectively or separately) can now be recognised as those by which chromatography functions, it seems fitting to group them, as KLEIN (1962) has done already, as the chromatographic processes.

PARTITION

A solute, if shaken in the presence of two immiscible liquids (phases), in which it is completely soluble, will distribute itself between the two phases. A solute may be introduced into a system involving two immiscible liquids in any of the following ways:

(1) addition in the solid form;

(2) dissolved in one liquid which is added to the other;

(3) dissolved in any or a pre-determined concentration in both liquids which are then mixed.

When equilibrium is reached, the ratio of the concentrations of the solute in the phases will be constant for that system (for the temperature prevailing and provided no change in molecular state occurs).

Thus

$$D_A = \frac{[A]_1}{[A]_2}$$

where D_A is a constant (the partition or distribution coefficient), $[A]_1$ and $[A]_2$ are the concentrations of the solute A in the phases *1* and *2* respectively.

For example, if a solution of molybdenum(VI) in 6 *M* HCl is shaken with an equal volume of diethyl ether, the molybdenum will distribute itself so that about 70% of it is contained in the diethyl ether while the rest remains in the acid phase (SANDELL, 1944).

Thus, numerically, for a given temperature,

$$D_A = \frac{30}{70} = 0.429$$

The presence of a second solute in the same system does not alter this relationship unless its presence alters the chemical composition or the molecular state of any species in the system. If a second solute *B* is distributed between the phases then

$$D_A = \frac{[A]_1}{[A]_2}$$

$$D_B = \frac{[B]_1}{[B]_2}$$

The extent of the separation of the two solutes is expressed as the separation factor (α) where

$$\alpha = \frac{D_A}{D_B} = \frac{[A]_1[B]_2}{[A]_2[B]_1}$$

If α approaches unity, the solutes will be distributed evenly and hence little separation will be achieved. Thus maximum separation depends on the existence (or choice) of phases that give values for the separation factor much greater than or much less than unity. Complete separation after one equilibration would result in all of solute *A* being in one phase and all of solute *B* being in the other. In this case

$$D_A = \infty$$

$$D_B = 0$$

or vice versa.

If complete separation is not achieved in the first equilibration, it may be achieved virtually by repeated equilibrations. Thus, if solute *A* separates mostly into phase *1*, further separation of *A* from *B* will be achieved in phase *2* by washing that phase repeatedly with new volumes of phase *1*. Conversely, further washings of phase *1* by new volumes of phase *2* will progressively eliminate solute *B* (and some of solute *A*, of course) from phase *1*.

For example, if a mixture of 1000 parts of A and 1000 parts of B is allowed to reach equilibrium in equal volumes of phases *1* and *2*—such that $D_A = 1$ and $D_B = 7$, then after the first equilibration, since

$$\frac{[A]_1}{[A]_2} = 1$$

and

$$\frac{[B]_1}{[B]_2} = 7$$

there will be

(*a*) in phase *1*: 500 parts of A and 875 parts of B,

(*b*) in phase *2*: 500 parts of A and 125 parts of B.

If, in the second equilibration, a new (equal) volume of phase *2* (shall we say phase 2_2) is mixed with the original phase *1*, there would be

(*a*) in phase *1*: 250 parts of A and 765 parts of B,

(*b*) in phase 2_2: 250 parts of A and 110 parts of B.

After a third such equilibration there will be

(*a*) in phase *1*: 125 parts of A and 669.4 parts of B,

(*b*) in phase 2_3: 125 parts of A and 95.6 parts of B.

The mathematical relationships may be expressed as follows: for solute A, the concentration remaining in phase *1* after the first equilibration will be

$${}^1/_2([A]_1 + [A]_2)$$

and after the second equilibration

$$^1/_4([A]_1 + [A]_2)$$

or

$$(^1/_2)^2([A]_1 + [A]_2)$$

and after the nth equilibration

$$(^1/_2)^n([A]_1 + [A]_2)$$

Likewise, for solute B, the concentration $[B_n]_1$ remaining in solvent *1* after n equilibrations will be

$$[B_n]_1 = (^7/_8)^n([B]_1 + [B]_2)$$

Further, if X_A is the original mass of A and X_B the original mass of B, then, after n equilibrations, the mass of A remaining in phases *1* and *2* will be respectively

$$X_A\left(\frac{[A]_1}{[A]_1 + [A]_2}\right)^n \quad \text{or} \quad X_A\left(\frac{[A]_1 D_A}{[A]_1(1 + D_A)}\right)^n$$

and

$$X_A\left(\frac{[A]_2}{[A]_1 + [A]_2}\right)^n \quad \text{or} \quad X_A\left(\frac{1}{1 + D_A}\right)^n$$

Similar expressions can be written for solute *B*.

By way of a further example, if, in such a system as described above, D_A had been 4 and D_B had been 0.25, then one can calculate that after the 20th equilibration 11.53 parts of *A* and $10.48 \cdot 10^{-12}$ parts of *B* would remain in solvent *1* and the 20th volume of solvent *2* would contain 2.88 parts of *A* and $41.88 \cdot 10^{-12}$ parts of *B*.

The examples illustrate that partition (or solvent extraction) requires large volumes of solvents and repeated manipulations. Even then complete separation is not achieved. Nevertheless, the method is often used, especially when the separation factor is large and when complete separation is not essential. The role of partition in chromatography will be considered after the other processes have been discussed.

DIFFUSION

The phenomenon of diffusion in gases and liquids is common and well-known. The diffusing particles may be either molecules (commonly in gases) or ions (commonly in liquids).

Gaseous diffusion is explained in terms of the kinetic theory. The very high velocities possessed by gas molecules point towards very rapid (almost instantaneous) diffussion. In fact, a much slower rate of diffusion is measurable and is expressed in terms of Graham's law, which states that the velocity of diffusion is inversely proportional to the square root of the density of the gas. The slower rate of diffusion is explained by the concept of frequent collisions between molecules and hence the limitation of free movement. Thus the *mean free path*, the mean distance over which a molecule moves between collisions, at N.T.P., is $6.43 \cdot 10^{-6}$ cm for oxygen and $11.2 \cdot 10^{-6}$ cm for hydrogen (FINDLAY, 1942, p. 59).

The whole process of gaseous diffusion is most complex, however, and, as yet, is not clearly understood (PARTINGTON, 1949, p.900).

Ionic diffusion in gases is restricted to very high temperatures. No consideration of this will be made since the temperature is above that of the normal geological processes.

Molecular diffusion in liquids is slower than in gases because the molecules are crowded more closely together and hence the mean free path is shorter.

Ionic diffusion in liquids proceeds because of differences in the nature, mobility and concentration of the ions (SHEEHAN, 1961, p.444). Thus the slight deviation from electroneutrality at an interface between solutions causes ions to migrate under the influence of a measurable potential difference.

Demonstrations of diffusion rates in liquids, such as the placing of a copper sulphate crystal at the base of a vertical column of water, tend to over-simplify complex situations. The influence of gravity and convection, for example, influence the rate in such a demonstration. The time required for observation of liquid diffusion is of the order 10^4 times that for gaseous diffusion. Thus JOST (1960, p.27) states "to cover an average distance of 1 cm takes a time of the order of magnitude of some seconds for molecules in gases at N.T.P., of some days for liquid solutions at room temperature and of between a day and 10^{12} years for such solids and temperatures where measurements have been possible".

A phenomenon often confused with diffusion in liquids is osmosis. GLASSTONE and LEWIS (1960, p.242) defines osmosis as "the spontaneous flow of solvent into a solution, or from a more dilute to a more concentrated solution, when the two liquids are separated from each other by a suitable membrane. Osmosis strictly refers to flow of solvent only; if there is movement of solute, in the opposite direction, the behaviour is then called diffusion".

The diffusion coefficient is defined as the number of moles of solute that diffuse across unit area in unit time under the influence of a concentration gradient of unity. The diffusion coefficient, D, can thus be expressed (according to Fick's law) as

$$\frac{dQ}{dt} = D \cdot A \left(\frac{-dc}{dx} \right)$$

where Q is the amount of substance in grams or moles,
t is the time interval,
A is the area over which the diffusion occurs,
$-dc$ is the decline in concentration per infinitesimal length in the direction of decline of concentration,
dx is the infinitesimal length.

The diffusion coefficient can be related also to absolute temperature (T), the gas constant (R) and Avagadro's number (N) in the one case and to the coefficient of viscosity (η) of the solvent in another.

Thus

$$D = \frac{RT}{Nf}$$

where f is the frictional force opposing a particle which is moving at a velocity of 1 cm/sec, and

$$D = \frac{RT}{N6\pi\eta r}$$

where r is the radius of a spherical particle. The derivations of these equations are given by DANIELS (1951, p. 531). Expressed in words, Fick's law states "that the rate of flow in any direction per unit area and per unit time is proportional to the concentration gradient in that direction" (TURNER and VERHOOGEN, 1951, p. 41). In ionic solutions, diffusion by the cation and anion are in the same direction, that is, in the direction of the concentration gradient. Thus diffusion maintains electric neutrality. This kind of movement is not restricted to diffusion by a single solute but operates also in binary, ternary and multiple mixtures of solutes in a common solvent. By contrast, cations and anions will move in opposite directions if the electric equilibrium is upset by the application of an external potential difference. This induced movement is best referred to as electromigration but is referred to by some authors as diffusion.

In solid diffusion, the concept of mean free path has no meaning since *most* particles in a solid merely oscillate about permanent or temporary equilibrium positions. In an ideal close-packed crystal at absolute zero all lattice spaces are filled. It can be shown (RAMBERG, 1952, p. 78) that, at all temperatures above absolute zero, holes or vacant sites will develop with or without evaporation. The possibility of ions diffusing through these holes is evident.

Diffusion coefficients in crystals will depend markedly on the crystallographic direction of flow. Anisotropy has very great effects on diffusion rates. Thus the self-diffusion of bismuth in bismuth at 275° C along the C axis is about a million times that at right angles to that axis (TURNER and VERHOOGEN, 1951, p. 406). Similarly, in quartz, in directions parallel to the C axis, there are open channels of such dimensions and thermodynamic stability that small ions can pass readily along them. Thus Ag^+ (ionic radius 1.26 Å), Li^+ (0.80 Å) and Na^+ (1.05 Å) may diffuse through such openings in quartz at 500° C but K^+ (1.35 Å) diffuses imperceptibly (RAMBERG, 1952, p. 81). A number of other silicate minerals, notably beryl, cordierite, feldspars, hornblendes and micas have fairly open lattices that permit diffusion of this kind.

The energy considerations of such migrations are not well established. However, evidence for such migration has been provided by BUERGER (1960) by order–disorder transitions and by the interpretation of ex-solution textures displayed by many minerals (EDWARDS, 1954).

When *lattice particles* migrate by the process of occupying vacant sites, resulting in fact in a migration of the hole, a change in concentration will occur. On the basis of energy considerations, this movement would tend towards concentration equilibrium and as such would be normal diffusion. If *extraneous particles* move interstitially and at no time occupy positions that belong to the normal lattice structure, their movement cannot lead to equalization of concentrations. Equalization can be achieved, however, by an interchange between particles on normal lattice sites and those on interstitial sites. Indeed, this kind of interchange is the very cause of diffusion in crystals. Thus JOST (1960, p. 144) states "...it may be concluded with absolute certainty without further experiments that ionic conduction and the corresponding diffusion process in crystals are due to lattice defects. Disorder, vacancies or interstitial particles must be responsible for these transport phenomena".

In solid ionic crystals, diffusion by different cations will be analogous to diffusion in ionic solutions. Since ionic size now plays a more important part, however, mobilities of the ions will not be the same. Thus, in the system

$$2AgI + Cu_2S = 2CuI + Ag_2S$$

the iodine and sulphur ions having large radii (2.16 and 1.84 Å respectively) will have small mobilities compared with those of Ag^+ (1.26 Å) and Cu^+ (0.96 Å). Therefore, the only appreciable result will be the interchange of silver and copper ions between the virtually immiscible iodide and sulphide phases.

Silicate compounds of the zeolite type have particularly open, wide-meshed $(Si,Al)O_2$ frameworks. The channels through the framework are sufficiently large to allow ions to pass freely. Further, when water molecules are expelled, the structure remains open. Diffusion of the interstitial type is readily effected along these channels without involving any mobility of the constituent particles of the crystals.

Diffusion in gels and colloids was the subject of spectacular experimentation and theorizing particularly in the first quarter of this century (LIESEGANG, 1913; HATSCHEK, 1922; ZSIGMONDY, 1917; BOYDELL, 1924). Molecular and ionic solutions in dilute gels diffuse at about the same rates as in similar molecular solutions. However, the rate of diffusion diminishes as the gel concentration increases or as the colloidal particle size increases. The diffusion of colloids in gels is negligible.

ADSORPTION

Adsorption is the phenomenon of substances being held at the surface of a liquid or solid in such a way that the concentration of the substance in the boundary region (surface or interface) is higher than in the interior of the contiguous phases. When the substance penetrates more or less uniformly through the liquid or solid, absorption is said to take place. Sorption is a general term that includes both adsorption and absorption or either. "The definition of adsorption rests upon measurements of concentration, and carries no implication of mechanism" (CASSIDY, 1951, p. 1). If the concentration in the boundary region is lower than in the interior, the substance is said to be negatively adsorbed. The substance adsorbed is called the *adsorptive* and the adsorbing phase is the *adsorbent*. The term *adsorbate* may refer to the adsorptive or to the adsorptive and adsorbent combined.

An interface exists between two or more phases. The thickness of an interface is of about molecular thickness (approximately a maximum of 100 Å). Interfaces may be designated according to the nature of the contiguous phases. Thus liquid–gas and liquid–liquid interfaces are clearly designated and, because they always tend to contract, are called *mobile interfaces*. Likewise, solid–gas, solid–liquid and solid–solid interfaces are said to be *immobile*, being influenced by the rigidity of the solid phase or phases.

All adsorptive processes result in a decrease in the total free energy of the system. Thus the processes are exothermic, the decrease in heat content being the heat of adsorption.

Some adsorptions, particularly gas–liquid adsorptions, are accompanied by low heats of adsorption of the order of 5–10 Kcal per mole. This is about the same order as heats of condensation. Hence these forces of adsorption are similar to the forces of cohesion of molecules in the liquid state or, in other words, to van der Waals forces. Adsorptions of this kind are called *physisorption*, *physical adsorption* or *van der Waals adsorption*. They are very common. Several molecular layers may be adsorbed, the outer layers being held more tenuously than the inner.

Other adsorptions, less common and observed mainly at moderately high temperatures, have heats of adsorption ranging from 10–100 Kcal per mole. These adsorptions are highly specific, generally irreversible and, partly or completely, involve forces of a chemical nature (ionic bonding, for example). Probably, due to the rapid diminution of chemical forces with distance, only single molecular layers are involved. This is called *chemisorption*.

Between these limits, and at moderately low temperatures, *activated adsorption* may occur at a much slower rate. There are no sharp boundaries between the three kinds of adsorption and borderline cases defy rigid classification. Some authors regard chemisorption and activated adsorption as synonymous.

Numerous factors favourable or unfavourable to adsorption are known to operate under the variety of conditions that may prevail. A very thorough treatment of adsorption in chromatography has been given by CASSIDY (1951). The summary of favourable and unfavourable factors which follows is mainly an extraction from Cassidy's work.

Factors favouring adsorption:

(1) The existence of interactions between solute and solvent in a liquid phase favouring escape of molecules from that phase to the interface. The process may be assisted by the presence of impurities.

(2) Temperature decrease is accompanied by increased adsorption.

(3) A higher degree of ionization of a solute in a liquid phase produces adsorbable ionic groups. Favourable pH conditions exert some influence here.

(4) The presence of favourable relationships between the chemical natures of the adsorptive and the adsorbent (such as favourable dipole relationships, for example, the dipoles of the adsorbent being able to polarize molecules about to be adsorbed).

(5) The movement in phase of electrons of fairly closely adjacent molecules about to be adsorbed is conducive to greater adsorption.

(6) The physical nature of the surface of the adsorbent influences adsorption. Favourable features are large surface area, corners and edges of crystals, plentiful distribution of sites at which adsorption can occur, roughness of surface, pore spaces, lattice channels and fissures to accommodate small and perhaps large molecules.

(7) Adsorption will only proceed, of course, if the adsorption sites are still vacant. Hence the presence of many vacant sites favours adsorption.

(8) The shape of the molecule to be adsorbed, and particularly the position of the polar groups in relation to the shape, influence

adsorption. Polar groups at the end are the most favourable positions.

Factors unfavourable to adsorption:

(1) If interactions between solute and solvent in the liquid phase oppose the escape of molecules from that phase to the interface, then adsorption will be diminished. Some impurities are unfavourable and reduce adsorption.

(2) Temperature increase is unfavourable.

(3) If the degree of ionization of the solute in the liquid phase produces unadsorbable ionic groups, adsorption will be weaker. Unfavourable pH conditions may aid the formation of such ionic groups.

(4) If unfavourable relationships between the chemical natures of adsorptive and adsorbent prevail, adsorption will be less. Such relationships as repulsion by like charges and molecules available for adsorption not polarizable by the dipoles of the adsorbent reduce adsorption.

(5) The mutual repulsion of molecules brought too close together is not favourable for adsorption.

(6) The unfavourable nature of the surface of the adsorbent is important. Such features as plane, crystalline, homogeneous surfaces, small surface areas, sparse distribution of favourable sites for adsorption and pores or similar openings already blocked by large adsorbed molecules are unfavourable.

(7) If the adsorption sites of the adsorbent are fully occupied, then further adsorption is not possible.

(8) The surrender of the three-dimensional freedom by the adsorbed molecule for the two-dimensional freedom of the interface, has an unfavourable influence on adsorption of other molecules.

(9) Competition for the adsorption sites in the interface by molecules of both the solute and solvent of the liquid phase leads to less adsorption of the desired molecule.

It seems unnecessary to emphasise the complex nature of adsorption. Its role in chromatography is equally complex. Perhaps it is fortunate that both of these processes go on without any pre-requisite human understanding.

ION EXCHANGE

Ion exchange has been defined (DOW CHEMICAL COMPANY, unknown date, p. 3) "as a reversible exchange of ions between a solid and a liquid in which there is no substantial change in the structure of the solid". Within the limits of the conditions under which they operate, ion exchangers are insoluble substances of such a nature that ions in their structure or in their interface are freely exchanged with ions from a solution in contact with the exchanger. Reversal of the exchange is readily achieved by even slight variations of the nature of the solution. Some ion exchangers operate only within a narrow pH range "as they peptize in alkaline solutions and dissolve in acid" (LEDERER and LEDERER, 1957, p. 72). Exchange of ions at the interface involves adsorption, while exchange from within their structure involves both sorption and diffusion. Cation exchangers exchange cations and anion exchangers exchange anions.

Natural ion exchangers were recognised first in soil chemistry (H. S. Thompson and J. T. Way, as stated by SALMON and HALE, 1959, p.1) the first observation being the power of the soil to absorb ammonia from ammonium sulphate accompanied by an equivalent yield of calcium. The first recognition of base preference was made when Way showed that ammonia could replace calcium whereas sodium could not (J. T. Way, as stated by SALMON and HALE, 1959, p. 1). Clays, humic acids and zeolites comprise the common natural ion exchangers.

Synthetic ion exchangers were firstly inorganic alumino silicates— synthetic zeolites. The organic synthetic ion exchange resins are prepared as large-molecule particles that exhibit "an elastic, three-dimensional hydrocarbon network to which is attached a large number of ionizable groups" (DOW CHEMICAL COMPANY, unknown date, p. 3). Variation of the hydrocarbon network changes the chemical nature of the exchanger in degree only. Modern exchangers are resistant to oxidation or reduction, to mechanical wear and fracture and to dissolution by common solvents. The ionizable groups can be varied to produce strong or weak cation exchangers and strong or weak anion exchangers. The strong cation exchangers have the sulfonic (–SO_3H) unit as the functional group. The weak cation exchangers contain carboxylic (–COOH) groups. The basic anion exchangers contain quaternary ammonium or tertiary amine groups.

The early organic networks were prepared by sulfonating carbonaceous materials (for example, coal or peat). Then came sulfonic acid cation exchangers prepared by the reaction of an aldehyde, a phenol and a sulfonic acid. More recently, a superior hydrocarbon network has been produced by the co-polymerization of styrene and divinylbenzene. The degree of crosslinkage is usually 8% of divinylbenzene but special resins with up to 16% divinylbenzene are available. Resins with small crosslinkage (e.g., less than 1%) take up many times their own weight of water and swell greatly. Tightly crosslinked resins (up to 40% divinylbenzene) inhibit absorption and hence ion exchange. The compromise at about 8% ensures the two basic requirements. "The resin must be sufficiently cross-linked to have but a negligible solubility..." but "must be sufficiently hydrophilic to permit diffusion of ions through the structure at a finite and usable rate" (KUNIN, 1958, p. 73). Further requirements are that the resins contain a sufficient number of accessible ionic exchange groups, that they are chemically stable during use and that the swollen resin is more dense than water.

The introduction of special groups into the hydrocarbon network produces special properties for specific tasks. If acidic groups (–SO_3H, –COOH, etc.) are attached to the network, cations from the solution will exchange with H^+ ions of the resin. If strongly basic groups are attached, anion exchange will be possible. For example, if a resin in the H^+ form is immersed in a sodium chloride solution, Na^+ ions will replace H^+ ions until an equilibrium condition is reached. Thus if R represents the resin network,

$$HR + Na^+ + Cl^- \rightleftharpoons NaR + H^+ + Cl^-$$

If the solution had been magnesium chloride, the reaction could have been expressed thus

$$2HR + Mg^{2+} + 2Cl^- \rightleftharpoons MgR_2 + 2H^+ + 2Cl^-$$

Some anions (in this example, Cl^- ions) have been shown to enter the resin phase (possibly by partition) but this does not effect the cation exchange which is strictly equivalent. Since both the reactions above are reversible, a sufficient increase in the H^+ ion concentration on the right-hand side will cause the reaction to proceed towards the left, H^+ ions displacing the Na^+ or Mg^{2+} ions from the resin. Hence, once separated from a solution by being fixed on a resin, cations can be recovered from the resin by strong acid treatment (generally about 10% HCl). This process is called *elution.*

Similarly with a strong basic resin in the $(OH)^-$ form, the following reactions can be achieved:

$$ROH + Na^+ + Cl^- \rightleftharpoons RCl + Na^+ + OH^-$$

and

$$2ROH + Mg^{2+} + 2Cl^- \rightleftharpoons 2RCl + Mg^{2+} + 2OH^-$$

Recovery of the sorbed Cl^- ions could be achieved by elution with a strong alkali solution.

The weakly basic resins have no $(OH)^-$ form. On treatment with alkali, they become only slightly ionized and consequently they perform satisfactorily only in near neutral or acid solutions. They exchange anions but are incapable of "splitting salts" as the reactions above indicate. A typical reaction would be the anion exchange achieved when the resin in the chloride form exchanges with the sulphate ion from a saline or acid solution.

$$2RCl + SO_4^{2-} + 2H^+ \leftrightharpoons R_2SO_4 + 2Cl^- + 2H^+$$

Elution with a strong chloride solution would release the SO_4^{2-} ions and regenerate the resin to the chloride form.

The capacity of an ion exchange resin may be expressed as the theoretical number of ionic sites per unit weight or volume of the resin. The measured (operating) capacity depends on the pre-treatment of the resin (for the determination), the nature of the solution and (to some extent) temperature. Dry weight capacities are usually expressed in mequiv./g and wet volume capacities in mequiv./ml. Thus Dowex 50W in the H^+ form is claimed to have a dry weight capacity of 5.0 mequiv./g. For example, if one equivalent of Ca^{2+} is 20 (the atomic weight 40 divided by the valency 2), then one gram of the resin would take up $5 \cdot 0.02$ or 0.1 gram of Ca^{2+} ions from a favourable solution.

Complex or large ions cannot be held effectively because their size prevents their entry into the resin channels.

Ions of higher valency, in general, are more strongly sorbed than those with lower. Apart from this, and especially when valencies are equal, most ion exchange resins exhibit a preference for the sorption of particular ions. Thus the apparent affinities of Zeo-Karb 215 are given in part as follows (PERMUTIT Co., unknown date, p.7):

$$Hg^{2+} < Zr^{4+} < Li^{+} < H^{+} < Na^{+} < K^{+}, \text{etc.}$$

This preference constitutes the *selectivity* of the resin. The selectivity coefficient *(Kc)* is analogous to an equilibrium constant. If a solution containing ions of *A* is brought into contact with a resin, already upon which ions of *B* are sorbed, both ions having the same valency,

$$(Kc)\frac{A}{B} = \frac{\text{concentration of } A \text{ in resin} \cdot \text{concentration of } B \text{ in solution}}{\text{concentration of } B \text{ in resin} \cdot \text{concentration of } A \text{ in solution}} = \frac{[A]_r\,[B]}{[A]\,[B]_r}$$

More complex relationships exist if the valencies of the ions are unequal and if the ions form complexes or slightly ionized molecules. For example, the sorption of Fe^{3+} ions by a cation exchanger was found to exceed the theoretical value. Some of the Fe^{3+} ions hydrolysed to form the complex FeO^{+}, which, because of its lower charge, was sorbed to a greater extent than the trivalent Fe^{3+} ion (PERMUTIT Co., unknown date, p. 11).

Ion exchange resins are sold in a water wet condition and should be kept moist. Drying tends to split the beads. Oven dried resins (110° C) can sorb about their own weight of water.

Ion exchange cellulose derivatives form a new group of ion exchangers. Some advantage is offered by their ability to swell, which leads to a rapid rate of exchange and the accommodation of larger molecules than is possible on resins. Faster elution of large molecules is also claimed by the manufacturers. Strongly acidic and weakly acidic cation exchangers and strongly basic and weakly basic anion exchangers are available in the form of floc, powder or paper. The total exchange capacities, as claimed, are comparable to those of the resin exchangers.

Ion exchange provides not only for remarkably controlled separation in chemistry but also its role in natural geological processes is becoming more evident.

PRECIPITATION CHROMATOGRAPHY

A form of chromatography, that has had some recent applications to geology (ZIEGLER, 1961; SPAIN et al., 1962) and that seems destined to provide further applications, is precipitation chromatography. The stationary phase contains ions which are capable of exchanging with others from the mobile phase. The reactions proceed strongly in a direction leading to the new precipitation and so are less reversible than those of ion exchange. For this reason, subsequent elution is not usually practised. The stationary phase, after the precipitation desired, is ashed or decomposed by oxidizing acids as a prelude to estimation by normal chemical or physical methods. For example, as a stationary phase, cadmium sulphide (alone or dispersed on cellulose) will precipitate silver sulphide from a mobile phase containing silver nitrate. Once precipitated, the silver sulphide is not easily eluted (it is eluted by gold nitrate solution) and so the estimation of silver is preceded by ashing the precipitate with or without its organic support.

Although somewhat similar, precipitation chromatography differs from ion exchange chromatography in a number of ways. In ion exchange the exchangeable ions of the stationary phase occupy certain predetermined positions in an open, constant organic network. In precipitation chromatography, a constant organic network may be absent and the whole stationary phase may undergo chemical decomposition. When the stationary phase is dispersed on organic supports (such as agar gel or cellulose powder), the process more closely resembles ion exchange. The more difficult reversibility of the process then becomes a matter of degree and, in fact, ordinary elution is possible.

Like ion exchange, precipitation chromatography lends itself to "batch" operation (DOW CHEMICAL COMPANY, unknown date, p. 20). The material normally forming the stationary phase of a column is stirred in a beaker (or tank) with the solution until equilibrium is reached. Thus the tasks of packing and extruding the column are eliminated.

Materials that have been used for precipitation chemistry include ammonium sulphide in agar gel, cadmium sulphide–cellulose (available from Macherey–Nagel and Co., Düren, Germany) and zinc sulphide mixed with Supercel pulp, at pH 4–6.8.

COMPLEX ION FORMATION

The concept of simple ionic dissociation alone cannot be maintained in the light of modern observations and theory. It can be shown (VAN ARKEL, 1956) that the energy relationships, the size and charge of ions and the shape of molecules contribute to the formation of complex ions. Thus, not only does water dissociate simply according to the equation

$$H_2O \rightleftharpoons H^+ + (OH)^-$$

but it is capable of forming the hydronium (or hydroxonium) ion.

$$2H_2O \rightleftharpoons OH^- + OH_3^+$$

Further the hydronium ion may become hydrated with one or more molecules of water forming more complex ions thus

$$OH_3^+ + H_2O \rightleftharpoons (OH_3 \cdot H_2O)^+$$
$$OH_3^+ + nH_2O \rightleftharpoons (OH_3 \cdot nH_2O)^+$$

Similarly, a salt like $CdCl_2$ can combine with Cl^- ions to produce such complex ions as $CdCl^+$, $CdCl_2^\circ$, $CdCl_3^-$ and $CdCl_4^{2-}$.

In crystals, the molecules, neutral atoms and charged ions assume an arrangement consistent with a minimum potential energy. The coordination of a particle is the number and arrangement of those particles which are its nearest neighbours. It will vary with different particles and different structures. Normally, a particle will occupy the coordination to which it is best suited because such an arrangement requires a minimum of energy. It follows that energy must be supplied if the particle is to be moved to a less favourable coordination. The coordination number of an element is the maximum number of particles (ions or molecules) which can be linked directly to a central ion. The coordination number will depend on the charge on the central ion and the valency of the neighbouring ions. Thus KETELAAR (1958, p. 54) gives the data contained in Table I.

TABLE I

MAXIMUM COORDINATION NUMBERS

Charge on central ion	1	2	3	4	5	6	7	8
Monovalent ions	2	4	5[1]	6	8	8	8	12
Divalent ions	1	2	3	4	4	5[1]	5[1]	6

[1] Occur rarely.

The size of the central ion and more particularly the relative sizes of central and neighbouring ions will influence the value of the coordination number.

The electrical neutrality of many oxides, halides and sulphides is lost by their reacting with further atoms of the same kind to form complex ions, as for example

$$SO_3 + O^{2-} \rightleftharpoons SO_4^{2-}$$
$$TeO_3 + 3O^{2-} \rightleftharpoons TeO_6^{6-}$$
$$SiF_4 + 2F^- \rightleftharpoons SiF_6^{2-}$$
$$SbCl_5 + Cl^- \rightleftharpoons SbCl_6^-$$
$$As_2S_5 + 3S^{2-} \rightleftharpoons 2AsS_4^{3-}$$

The complex ions may then react with an ion or ions of appropriate opposite charge to form salts as the following which ionize as shown:

$$CsSbF_6 \rightleftharpoons Cs^+ + (SbF_6)^-$$
$$(NH_4)_4P_4O_{12} \rightleftharpoons 4(NH_4)^+ + (P_4O_{12})^{4-}$$

Similarly, the SiO_4 groups, familiar to most geologists, form either salts or complex ions (and also salts of polysilicic acid). Thus, in forsterite, Mg_2SiO_4, the ionic structure may be represented as follows:

$$2Mg^{2+} + \left[\begin{matrix} & O & \\ O & Si & O \\ & O & \end{matrix}\right]^{4-}$$

and

$$Ti^{4+} + Mg^{2+} + \left[\begin{array}{c} \mathrm{O\ O} \\ \mathrm{Si} \\ \mathrm{O\ O} \\ \mathrm{O\ Si\ O\quad Si\ O} \\ \mathrm{O\qquad O} \end{array}\right]^{6-}$$

In beryl, six SiO_4 groups in a closed ring form a complex ion with twelve negative charges (VAN ARKEL, 1956, p. 142).

Many complex organic ions form compounds with inorganic ions. These form the basis of many "spot tests" and gravimetric analyses. Thus the compound 8-hydroxyquinoline, which forms complexes with more than thirty elements, under suitable conditions, forms a monovalent $(C_9H_6ON)^-$ ion.

Some chromatographic behaviour and some methods of identification of chromatographic "spots" will be explained later by recourse to the concept of complexes and complex formation.

THE ROLE OF THE CHROMATOGRAPHIC PROCESSES IN CHROMATOGRAPHY

Fundamentally, the three main kinds of chromatography all have certain features in common:

(1) The support which consists of a column or a strip of solid material which is insoluble in the solvents to be introduced into the system.

(2) A suitable solvent forming a solution of the solute(s) to be separated. The solutes are sorbed from the solution onto or into the support.

(3) Another suitable solvent which is made to move over the support and the sorbed solute(s). This mobile phase causes the solutes to move (or not to move) so that they become concentrated in zones, the process being known as development. The mobile phase is also known as the eluting solution or the eluant.

If the support is a known adsorbent, such as alumina, magnesia or charcoal, adsorption is thought to play a major role in the chromatographic process. This process is then known as *adsorption chromatography*. If the support is a substance which absorbs a solvent (usually water) into its interstices, the absorbed solvent becomes a stationary phase. Such supports as silica gel, cellulose or rubber act in this way. The solute mostly passes from the solution to the stationary phase mainly by partition. Subsequently, elution by the mobile phase separates the solutes again mainly by partition. Hence, this kind of chromatography is called *partition chromatography*. If the support is an ion exchanger (synthetic or natural) it is capable of sorbing the solutes from the solution mainly by ion exchange. Elution is later achieved by ion exchange. This, then, constitutes *ion exchange chromatography*.

Adsorption chromatography

Almost any substance exhibiting a large surface area per unit volume can act as an adsorbent provided it does not react with or dissolve in the solvents to which it is to be exposed. Commonly, the substances that fulfil these conditions are in the form of powders or fibres. Strong acids and bases (because of their reactivity), most organic adsorbents (because of the difficulty in drying and regeneration) and dark coloured adsorbents (because of the difficulty of recognising coloured zones on the columns) are not favoured as adsorbents in chromatography. The adsorbent properties are influenced by the quality of the adsorbent and the method of its preparation. The latter often influences particle size and the degree of hydration. High temperature treatment may reduce the surface area of the adsorbent. For example, the surface area of alumina is 15% smaller when heated to 734°C than when heated to 528°C (KRIEGER, 1941).

The processes of adsorption chromatography can be explained more readily by consideration of a few examples. If an aqueous solution (0.2 ml), containing V^{5+}, Cr^{3+} and Co^{2+} ions, is introduced at the top of an alumina column, development may be effected in three ways, each involving adsorption as the main process.

As an example of *elution development*, if 50 ml of water are added to the top of the column, the three ions will form coloured zones and soon it will be apparent that the yellow V^{5+} zone remains almost stationary at the top, the greenish Cr^{3+} zone takes up a position below the V^{5+} zone and also remains almost stationary. The pinkish Co^{2+} zone will move down the column. After 13 ml of water pass out of the column, the Co^{2+} will appear in the effluent. After a total of 30 ml of effluent has flowed, all the Co^{2+} will be eluted from the column. Further large quantities of water cause the other zones to move only slightly.

The same column can now be used to illustrate the *liquid* (or *flowing*) *chromatogram* (ZECHMEISTER and CHOLNOKY, 1950, p. 76). If 50 ml of 0.2 *N* HCl are now introduced, the V^{5+} zone moves very slowly but the Cr^{3+} zone moves quickly down the column and is eluted by the time 50 ml of effluent have flowed. If 50 ml of 0.4 *N* HCl are now added, the V^{5+} zone will move down the column and will be eluted after 30 ml have flowed. Thus elution is achieved by use of a series of progressively stronger liquid phases which constitute the liquid chromatogram of Reichstein (as quoted by LEDERER and LEDERER, 1957, p. 30).

To illustrate *frontal analysis*, the author obtained the following results when 50 ml of a solution of the same ions in 0.4 *N* HCl were introduced to the column:

Effluent 0–13 ml contains solvent only.

Effluent 13–20 ml contains solvent + Co^{2+}.

Effluent 20–32 ml contains solvent + Co^{2+} + Cr^{3+}.

Effluent 32–50 ml contains solvent + Co^{2+} + Cr^{3+} + V^{5+}.

It will be noted that the solutes are not separated by this technique. Nevertheless the results, especially if the concentrations of the ions in the effluent are measured precisely, throw light on the nature of the original solution. A. Tiselius, using complex equipment, showed that in frontal analysis the variations in concentration of the effluent could be interpreted to indicate the character of solutes too complex to be resolved by other methods (BRIMLEY and BARRET, 1954, p. 21).

A fourth method of column development *(displacement development)* is considered to be essentially an ion exchange process and will be mentioned in that section.

Table II sets out the distinctive features of each of the development methods already described.

TABLE II

COLUMN DEVELOPMENT

Elution development	*Liquid chromatogram*	*Frontal analysis*
(*1*) Solutes in minimum of solvent added to column.	(*1*) Solutes in minimum of solvent added to column.	(*1*) Solutes in relatively large volume of solvent added to column.
(*2*) Pure solvent added.	(*2*) A series of solvents of increasing eluting power added.	(*2*) No further solvent added.
(*3*) Effluent contains in order: (*a*) Pure solvent. (*b*) Solutes in order of increasing absorbability with or without overlap. (*c*) Pure solvent.	(*3*) Effluent contains in order: (*a*) Some of the first solvent. (*b*) A series of effluents each containing a different solute. (*c*) Some of the last solvent.	(*3*) Effluent contains in order: (*a*) Pure solvent. (*b*) Solution of the least absorbed solute increasing in concentration to that of original solution. (*c*) Mixed solution of first solute at maximum concentration together with solution of second solute which increases to maximum concentration, and so on for the other solutes.

It should be clear that adsorption chromatography involves mainly the process of adsorption. Solute ions enter the interface of the adsorbent when their concentration in the solution is high, the solvent (liquid phase), almost free of solute ions, passes down the column. When the concentration of solute ions in the interface increases or their concentration in the solution decreases, solute ions pass out of the interface into the liquid phase. As the solution moves down the column the solute ions are adsorbed and desorbed repeatedly.

Some diffusion must occur within the liquid phase to carry the ions towards the interface where adsorption occurs.

Alumina, a common column adsorbent, for example, is strongly polar in nature and adsorption on it occurs through chemisorption. However, the presence of minute quantities of calcium on the column increases the adsorption capacity. This seems to indicate that ion exchange occurs too.

Development by the liquid chromatogram also implies some preference in the exchange of ions at the interface. Whether or not this preference is any more than a displacement of early adsorbed ions of low affinity by the late arrivals of ions of higher affinity is difficult to determine. The net result of such a process, however, is the separation of ions into zones.

Complete understanding of the processes involved in such columns has not yet been reached. LEDERER and LEDERER (1957, p. 429) state "Adsorption columns in use for inorganic separations function mainly as fractional precipitation, hydrolysis or complexing columns rather than true adsorption columns and, in some cases, the mechanism of separation is not clear at all".

The separation of inorganic cations, from aqueous solutions on alumina, depends on the valency of the ions, the pH of the solution and the affinity of the ions for adsorption. Higher valency ions are more strongly adsorbed than those of lower valency. From solutions with low pH, ions are less strongly adsorbed than from neutral or alkaline solutions. The amphoteric nature of the alumina tends to preserve a neutrality. Schwab and Jockers (see ZECHMEISTER and CHOLNOKY, 1950, p. 306)) have arranged the common ions in descending order of adsorption from nearly neutral solutions as follows:

$$As^{3+}; Sb^{3+}; Bi^{3+}; Cr^{3+}; Fe^{3+}; Hg^{2+}; UO_2^{2+}; Pb^{2+}; Ag^{+}; Zn^{2+}; Co^{2+}; Ni^{2+}; Cd^{2+}; Fe^{2+}; Tl^{+}; Mn^{2+}.$$

From dilutely ammoniacal solutions, the order for some ions is changed to

$$Co^{2+}; Zn^{2+}; Cd^{2+} Cu^{2+}; Ni^{2+}; Ag^{+}.$$

Thus Co^{2+} and Ni^{2+}, although inseparable from a neutral solution, may be separated easily from each other in an ammoniacal one. The presence of organic anions, which form complexes with inorganic cations, also alters the order of adsorption. Hydroxyquinoline and the tartrate ion are effective in this way (ZECHMEISTER and CHOLNOKY, 1950, p. 313).

Partition chromatography

Partition chromatography can be effected in columns, where finely divided material forms the support, and where this material is itself supported in a glass tube fitted with a porous pad or sintered disc. Alternatively, partition chromatography can be carried out on sheets of filter paper which act as self-supporting supports.

In column partition chromatography, supports such as silica gel, cellulose, starch and kieselguhr (diatomaceous earth) are used. The water which is arranged to saturate the support becomes the stationary phase. Sometimes buffer solutions such as tartaric acid and citric acid (ELLINGTON and STANLEY, 1955) are used instead of water as the stationary phase. The solvents that can be used are numerous. The prime condition is that they are immiscible with the stationary phase and the main restriction that they are not extremely volatile. The degree of this restriction is indicated by the successful use of ether as a solvent.

In paper partition chromatography, the support is the paper. The nature of the stationary phase is a matter of some controversy (LEDERER and LEDERER, 1957, pp. 115–116). It is fairly clear that the stationary phase is not just water held in the paper. An obscure water–cellulose relationship, in which, perhaps, two partition phases exist (A. P. J. Martin, as quoted by LEDERER and LEDERER, 1957, p. 115), seems likely. At this stage of our knowledge, it might be best to use generalities. The nature of the water–paper relationship may be regarded as a "water–cellulose complex" (HANES and ISHERWOOD, 1940). The solvents (mobile phases) move either up or down the paper. By changing the position of the paper, two-dimensional movement of the solvents across the paper is possible.

It has been shown at the beginning of this chapter that partition conducted in separating funnels or similar apparatus requires large volumes of solvent and repeated manipulations. In column and paper partition chromatography a continuous flow of the mobile phase over the stationary phase is achieved. Thus the discontinuous steps of the separating funnel are replaced by a continuous partition, which, in fact, amounts to an infinitely large number of discrete steps by infinitely small volumes of the mobile phase. Since one phase is stationary and the other mobile, the relative movement is counter-current. Partition chromatography is a counter-current solvent extraction process.

In a column or on paper, the relative movements of the solvent and the zone (solute) are expressed by the R_F value. Thus

$$R_F = \frac{\text{distance travelled by zone}}{\text{distance travelled by liquid front}}$$

An example of column partition chromatography is the method of KEMBER and WELLS (1951) for the extraction and determination of gold. A solution of the sample (containing gold and other metals) is placed on a cellulose column (7.5 cm $\times$ 2 cm diameter) with absorbed water forming the stationary phase. As solvent, 100 ml of a mixture of ethyl acetate containing 2% (volume for volume) nitric acid (s.g. 1.42) and 3% (v/v) water are used. The gold is eluted completely from the column by the volume of solvent used, Cu^{2+} moves 2 cm, Pt^{4+}, Pd^{2+}, Ir^{4+}, Rh^{3+}, Ru^{+} move 1 cm and Ni^{2+}, Co^{2+}, Mn^{2+}, Pb^{2+} and Mg^{2+} do not move.

Apart from partition, which is the dominant process in column partition, other chromatographic processes are involved. We have seen earlier in this chapter that absorption is accompanied by diffusion. Likewise, the concentration gradients, set up when partition occurs at the mobile phase–stationary phase interface, provide the conditions that promote diffusion of solute through the stationary phase to the interface and through the mobile phase from the interface.

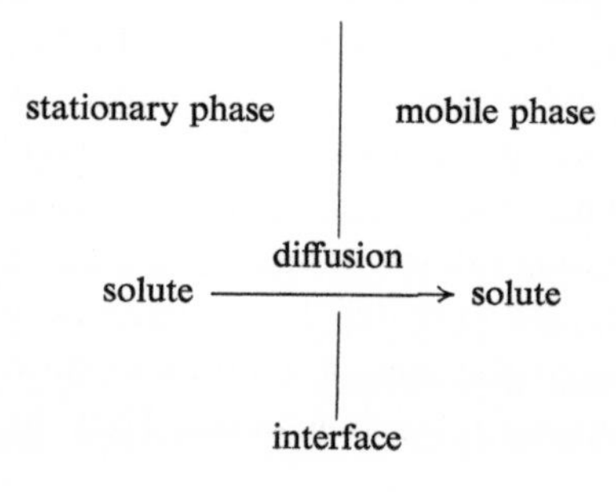

PICKERING (1959, p. 114) has shown that in the chromatographic column, cellulose is not inert (as one expects of the support as defined) but has a measurable adsorptive capacity.

CETINI and RICCA (1955–1956) have measured the varying adsorption capacities of silica gels prepared in a number of different ways. Likewise, the partition behaviour of silica gel in experiment and the calculated behaviour in theory are separated by disparities that are explained by some degree of adsorption by the gel (BASOLO et al., 1963, p. 64). Thus, in the column, silica gel cannot be regarded as inert. The method of preparation influences particle size and degree of hydration. These, in turn, influence the relative importance of adsorption in the column processes.

Column chromatography has certain advantages over paper chromatography. The dimensions of the column can be adjusted to accommodate any desired load of solute and the rate of flow of effluent can be varied within limits. The column lends itself to a variety of development techniques which can be used in rotation on the same column. The column can be used repeatedly under conditions that favour reproducibility.

Compared with column chromatography, paper chromatography provides opportunities for a greater variation of materials (such as the grade of paper). It permits the use of extremely small quantities of the sample and it provides permanent or semi-permanent records of the analysis. The possibility of two-directional development is another important advantage.

Two distinct vertical techniques (and at least one horizontal) are used in paper chromatography. The solvent may be allowed to move vertically by capillarity up the paper which has its lower edge immersed in the solvent, the solutes having been introduced first at a position just above the solvent level. This is the *ascending development*. Alternatively, the solvent may be introduced at the top of the paper (near which the solutes have been added) and allowed to descend under the influence of gravity. This is *descending development*. In *radial (circular) development*, the paper is held horizontally and the solutes and then the solvent are added at the centre. Rotation may be introduced to bring centrifugal force into play.

The ascending development calls for the simplest of arrangements but obviously the movement of the solvent is limited to the top of the paper. Travel against gravity is slow, 2–3 cm/h being favoured (MILTON and WATERS, 1955, p. 5).

Descending development is faster and the distance moved by the solvent can be increased by allowing it to drip off the lower end of the paper. This latter is an advantage for use with ions of low R_F values.

The principal advantage of circular development is that the ions are arranged in circular zones instead of spots. As a zone spreads from the centre (and assumes a larger circumference) its concentration is reduced.

The solvents (mobile phases) used in paper chromatography are numerous and varied (see Table VIII). Some separations have been effected using water as the solvent (BASOLO et al., 1963). More often the solvent is a single phase mixture of an organic solvent (such as butanol or ethanol) and a mineral acid (such as hydrochloric or hydrofluoric acid).

In all three techniques, the solute either moves with the solvent front, lags behind it to some extent, or remains unmoved on the paper. The actual mechanism by which the solutes are moved is not yet clearly understood. In some cases it appears that the ions form complexes with the acid components of the solvent mixture. However, for ions such as Na^+ and K^+, which do not form complexes with acids, some other process must achieve the differing R_F values obtained experimentally. The cellulose of the paper itself may form complexes (SMALES and WAGER, 1960, p.421).

The essentials of paper chromatography, so far as the geologist is concerned, may be stated as follows. A drop or less of solute (for example, solution of a mineral) is "chromatographed" by allowing solvents to move across the chromatography paper past the point of application of the solute. The metal ions move across the paper usually according to their ability to form complexes. The movement of the ion proportional to that of the solvent is called the R_F value of the ion in that solvent:

$$R_F = \frac{\text{distance moved by the spot}}{\text{distance moved by the solvent front}}$$

Some ions in some solvents form coloured spots which can be marked on the chromatogram as soon as it is taken out of the vessel. Other ions, being colourless on the paper, need to be treated with reagents (by spraying or by dipping) to produce either coloured or fluorescent spots. Radioactive ions (natural or irradiated) can be located by radiometric means.

Complex formation, adsorption, partition and ion exchange all seem to play some part in the complex mechanism of paper chromatography.

Ion exchange chromatography

Like partition chromatography, ion exchange chromatography can be effected either on columns or on paper impregnated with ion exchangers. "Batch" operation is a third method of using ion exchangers.

Ion exchange columns are prepared usually from small beads of synthetic ion exchange resins. Natural ion exchangers (such as the zeolites) and the newly introduced ion exchange crystals are alternative column materials. Ion exchange impregnated paper combines the smallscale analysis (capable on paper) with the diversity of nature and of capacity of synthetic ion exchangers.

At a time when only natural ion exchangers were being used, ion exchange was regarded as an adsorption process. The impervious nature of the mineral surfaces seemed to limit ion exchange to the interfaces. The advent of open network synthetic ion exchangers permitted the phenomenon of ion exchange to take place within and throughout the ion exchange substance. The phenomenon is regarded now as one involving both adsorption and absorption — that is, sorption.

Ion exchange column chromatography involves the sorption of solute ions onto the exchanger and, subsequently, the successive removal of them from the exchanger. The removal of the ions is accomplished in a number of ways. By *displacement development*, a less strongly sorbed ion is removed from the exchanger by a more strongly sorbed one. For example, if a 10% solution of sodium chloride is passed through a column of a cation exchanger in the hydrogen form, the sodium ions will displace the hydrogen ions and become sorbed on the resin. If a 10% solution of calcium chloride is passed through the column, the calcium ions will displace the sodium ions on the resin. The effluent will contain sodium chloride. *Frontal analysis* is effected in the same way as in other column separations. If a constant composition solvent is introduced to the top of the column, separation of the sorbed ions is achieved, to some extent, because some ions are exchanged more readily than others. The former move down the column more rapidly to form overlapping zones, which, upon leaving the column, produce a variable composition effluent. Likewise, *elution development* takes place when sorbed ions of several metals are removed from the exchanger successively by the passage of solvents of progressively increasing concentration.

To permit ion exchange paper chromatography, papers have been prepared in a number of ways, for example, by the formation of exchangeable carboxyl groups on the paper, The production of paper impregnated with ion exchange resin, which exhibits similar behaviour to resin columns (M. LEDERER, 1955; LEDERER and KERTES, 1956) was a significant advance. Now, a range of ion exchange papers is available from the leading manufacturers.

The complex nature of the role of a number of chromatographic processes in ion exchange columns is by no means clear. The introduction of a complex support (the paper) in ion exchange paper chromatography makes the total process more complex.

A better understanding of the role of partition, diffusion, adsorption, complex ion formation and ion exchange will come to the reader when the techniques have been mastered and the applications understood.

CHAPTER 3

Chromatographic Techniques

GENERAL

Usually, chromatography columns are packed with great care in glass tubes, often of the burette type. The glass itself confines the adsorbent, which is, however, supported from below by a porous pad or a sintered or porous plate fixed into the glass tube. A wide choice of shape and size of chromatography columns is available. Refinements have been designed to permit the flow of solvents through the column under reduced pressure at the lower end or increased pressure at the top of the tube. The vertical direction of flow of the solvent can be upwards or downwards. Often inlet and outlet reservoirs are fitted to permit constant flow rates over long periods. Adequate treatments of this subject are to be found in numerous textbooks (LEDERER and LEDERER, 1957; MILTON and WATERS, 1955; BRIMLEY and BARRET, 1954; STRAIN, 1942). Glassware for chromatography, with standard groundglass connections (cones and sockets), is very convenient and lends itself to variety in assembly. For field work, polythene and colourless plastic tubing form suitable columns (CANNEY and HAWKINS, 1960; CARRIT, 1953; COULOMB and GOLDSTEIN, 1956a). The heights of columns are usually at least ten times the diameters. For ion exchange columns the length may be much greater. For small masses of solute in the test solutions, shorter columns are often sufficiently efficient.

PREPARATION OF TEST SOLUTIONS

In all chemical analytical methods, an initial problem is to convert the material to be analysed into a form suitable to the analytical method. In chromatography, the material is required in solution.

In the case of natural waters and oils, the test material is often ready for analysis. However, solid test material (rocks and minerals) must be taken into solution.

Acid dissolution

Many minerals are soluble in the common mineral acids or in aqua regia. DANA (1932, p.360) lists the following minerals which are insoluble in acids: corundum, spinel, chromite, diaspore, rutile, cassiterite, quartz, cerargyrite, silicates, titanites, tantalates, niobates, sulphates (barite, celestite), phosphates (xenotime, lazulite, childrenite, amblygonite) and boracite.

Silver, lead and mercury minerals give insoluble chlorides with hydrochloric acid but go into solution in nitric acid. On the other hand, in the presence of nitric acid, the oxides of tin, arsenic and antimony are precipitated. Likewise the oxides of titanium, wolfram, molybdenum and vanadium are prone to precipitation during the decomposition of minerals containing them.

The problem of dissolution of opaque ore minerals, as far as is necessary for the identification of the constituent metals by paper chromatography, has been investigated by the present author (RITCHIE, 1962a). In view of the minute quantity of test solution required for paper chromatographic analysis, and hence, in view of the small quantities of sample and acids involved, it was possible to obtain a near-universal method of dissolution of ore minerals by the following procedure:

(a) Grind finely about 0.05 g of the mineral and transfer to a small test tube.

(b) Add 10 *N* HCl (about 10 drops) and, if necessary, warm. Many minerals will dissolve in this acid and in these cases no further treatment is required (see Table III). Retain the solution for analysis.

(c) If complete dissolution is not achieved, add 15 *N* HNO_3 (1 drop) and warm. Most ore minerals which are insoluble in HCl will be found to be soluble in this "nitric-acid-weak" aqua regia. Some minerals under these conditions precipitate sulphur, the

TABLE III

ORE MINERALS SOLUBLE IN HYDROCHLORIC ACID

Altaite	Galena	Pyrargyrite
Antimony (native)	Goethite	Pyrolusite
Argentite	Hessite	Pyrrhotite
Bismuth (native)	Hausmannite	Sphalerite
Bismuthinite	Jamesonite	Stibnite
Boulangerite	Lead (native)	Stromeyerite
Chalcocite	Magnetite	Tenorite
Copper (native)	Manganite	Tetradymite
Cuprite	Petzite	Zinkenite
Dyscrasite	Polybasite	
Franklinite	Psilomelane	

presence of which is recognised after short experience. Precipitated sulphur is ignored henceforth in these directions. Some insoluble chlorides might be precipitated by the hydrochloric acid or the chlorine from the aqua regia.

(d) Boil to drive off hydrogen chloride and chlorine and to reduce the bulk of liquid to a few drops. Add 15 N HNO_3 (one drop at a time) and warm. The insoluble chlorides will be taken into solution. However, oxides of tin, arsenic and antimony may be precipitated by the nitric acid. At this stage, the maximum dissolution by the hydrochloric acid-nitric acid-aqua regia treatment has been achieved (see Table IV).

TABLE IV

ORE MINERALS SOLUBLE IN NITRIC ACID
(Not including minerals appearing in Table III)

Arsenopyrite	Freibergite	Proustite
Bornite	Glaucodot	Pyrite
Bravoite	Linnaeite	Siegenite
Breithauptite	Loellingite–Safflorite	Skutterudite
Calaverite	Marcasite	Stannite
Chalcopyrite	Meneghinite	Sylvanite
Cobaltite	Millerite	Tennantite
Covellite	Niccolite	Tetrahedrite
Domeykite	Pentlandite	Wurtzite
Enargite		

(e) If a residue remains, decant off solution and retain for analysis.

(f) Wash residue with water and transfer to 10 ml polythene microbeaker. Add 48% HF (3 ml) and heat on water bath until only a paste remains.

(g) Add 48 HF% (3 drops) and 15 *N* HNO_3 (1 drop). Warm. The method of SHAPIRO and BRANNOCK (1956, p. 32, for solution B) modified to suit the smaller samples may be substituted for this step. Chromatograph the solution.

(h) If a residue remains, an alkaline fusion of a fresh sample is indicated.

Alkaline fusion

In view of the small quantity of sample and solution required, the alkaline fusion may be carried out in an alkali bead supported on a loop of nichrome wire. A mixture $Na_2CO_3:KNO_3 = 19:1$ is most satisfactory. In the laboratory, a bunsen burner provides sufficient heat. In the field, the flame of a candle or of a kerosene lamp may be used in conjunction with a blowpipe. Alternatively, a primus flame or a "portogas" flame may be used. When fusion is complete, the bead is cooled, crushed and dissolved in about 1 ml of warm water. If residues appear add 15 *N* NHO_3 (a few drops). The resulting solution is evaporated to about 0.1 ml and chromatographed. Alternatively, if the reduction in volume induces precipitation, a larger volume (1 or 2 ml) can be preserved and an aliquot taken.

Decomposition by sodium peroxide has been well-known to analysts for a long time, but its use has been limited in the past because of its corrosive action on the metals of the crucibles (e.g., platinum, gold or nickel). Three serious disadvantages were the high cost of crucible replacement, the presence of corroded metal in the analysis and the loss to the corroded crucible of trace metals from the sample.

It has now been established (DOBSON, 1962; BELCHER, 1963; RAFTER, 1950) that decomposition by sodium peroxide of even the most refractory minerals can be achieved between temperatures of

260° C and 500° C, that corrosion of the metals already mentioned is slight in that temperature range and that pure zirconium vessels suffer negligible corrosion. The availability of pure zirconium vessels (at high, but not unreasonably high, prices) has coincided with these discoveries.

The method readily decomposes silicates, oxides, sulphates, tantalates, niobates and phosphates (see Table V). Sulphides react

TABLE V

PRODUCTS OF SODIUM PEROXIDE FUSIONS

Minerals and formulae	*Water soluble products*
corundum, Al_2O_3	sodium aluminate, $Na(AlO_2)$
chromite, $FeCr_2O_4$	sodium chromate, Na_2CrO_4; sodium ferrate Na_3FeO_3
rutile, TiO_2	sodium titanate, Na_4TiO_4
cassiterite, SnO_2	sodium stannate, Na_2SnO_3
tantalite, $(Fe,Mn)Ta_2O_6$	sodium ferrate, Na_3FeO_3; sodium tantalate, $Na_2Ta_2O_6$; sodium manganate, Na_2MnO_4
cerargyrite, AgCl	silver complex
sphene, $CaTiSiO_5$	sodium titanate Na_4TiO_4; sodium silicate, Na_2SiO_3; calcium hydroxide, $Ca(OH)_2$
amblygonite, $(Li,Na)Al(PO_4)(F,OH)$	sodium aluminate, $Na(AlO_2)$; lithium hydroxide, LiOH
eskebornite, Fe Se, Cu Se	sodium ferrate, Na_3FeO_3; copper hydroxide; sodium selenate, Na_2SeO_4

exothermally with sodium peroxide and often result in high crucible temperature with consequent corrosion. However, in view of the small quantity of sample required for paper chromatography and of the low temperature for decomposition, sulphide and sulpho-salt decomposition can be carried out safely with reasonable care. Thus a universal method for the decomposition of all minerals presents itself. For chromatographic work (in both the laboratory and the field) the following procedure is recommended.

Zirconium crucibles of 3 to 5 ml capacity can be bought or pressed from sheets of high purity zirconium of thickness ranging from 0.0035–0.03 inch. The sheet initially should have a bright polish and this should be preserved for as long as possible. Since the metal can be oxidised by unequal and excessive heat to a white oxide, it is important to take precautions to prevent the break-down of this surface. A simple heat-exchanger (see Fig. 1) can be made by bedding the crucible to its brim in A.R. (analytical reagent) quality anhydrous sodium carbonate contained in a 30 ml dish, preferably of nickel or platinum. A small watch-glass serves as a cover.

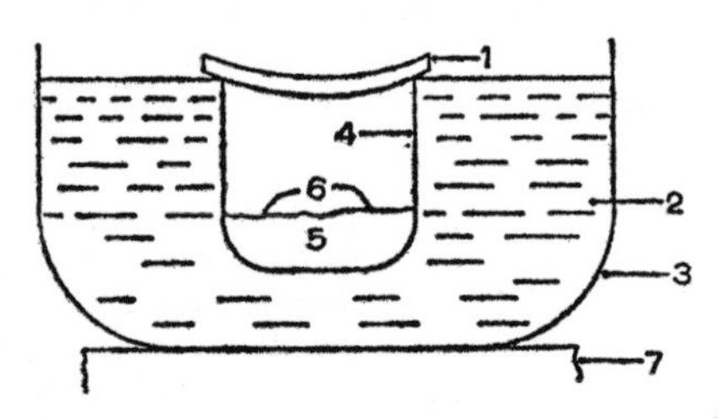

Fig.1. Sodium peroxide fusion. Zirconium crucible in simple heat-exchanger. *1* = watchglass cover. *2* = sodium carbonate. *3* = nickel, platinum or iron dish. *4* = zirconium crucible. *5* = sodium peroxide. *6* = mineral. *7* = hot-plate.

In the laboratory, the exchanger should be heated by an electric hot-plate at temperatures below 450° C. At a base camp, a steel plate heated from below by "portogas" or a primus could be used.

Finely grind the sample and weigh out about 0.05 g. Hard minerals (topaz, sapphire, tourmaline, quartz) are best pulverized in a tungsten carbide percussion mortar. Softer minerals (magnetite, feldspars, etc.) may be ground in an agate mortar. Place about 1 g of A.R. grade Na_2O_2 in the crucible and add the sample on top. Heat the crucible through the exchanger. Because some reactions are exothermic, heating should be increased gently until the reaction begins, whereupon no further heat may be required. It is important to realize that the reaction may begin *before* the mixture fuses. Decomposition is usually achieved with these quantities in 1–5 minutes. The crucible is cooled rapidly by dropping it into 5–10 ml of cold water. The sintered cake usually dissolves readily in the alkaline solution so formed.

Stir well, allow any excess Na_2O_2 to settle and decant off the solution of the metal(s). Wash residue with 2 ml water and decant. Reduce volume of the decanted liquids on a water-bath to 1 ml or less. This solution is then ready to be chromatographed. If precipitates appear, take them into solution with 15 *N* HNO_3 (a few drops).

It is to be expected that a general method, for the variety of material that is encountered in rocks and minerals, will be somewhat complex. For many minerals and some rocks, a simpler method of dissolution will be available. No doubt, readers will choose a specific method to suit their material. For unidentified material the general method commends itself.

ADSORPTION COLUMNS

The techniques for the operation of adsorption columns have been developed mainly for organic separations. In general, however, they hold for inorganic ions even though it has become clear that the chromatographic behaviour of inorganic ions in these circumstances is due to an hydrolysis process rather than the *real* adsorption of organic compounds (M. Lederer, personal communication, 1963). SMITH (1953) draws some distinctions between organic and inorganic column chromatography.

Packing

The common objective of all packing techniques is to produce a column through which liquid flow will be uniform and in which the substances to be separated will form flat horizontal zones. Faults to be avoided are channelling, gas bubbles and graded sedimentation, the latter being generally less serious than the other two.

A dry method, based on the addition of the adsorbent in small quantities, each of which is tamped down firmly and uniformly with a convenient rod, has been used extensively. Lack of uniformity of compaction and a prevalence of bubbles are likely faults with this technique. Suction applied at the lower end of the tube assists in achieving a better result. Alternatively, the adsorbent may be poured into the tube, in small quantities at a time, as a "slurry". After each addition, the column is tamped to aid compaction and to prevent graded sedimentation. The tap at the lower end of the tube is left open and so the water content is slowly reduced at the top of the column during preparation. During and after packing, the column should never be allowed to dry out lest irreversible shrinkage of the adsorbent occurs and air bubbles are introduced. Some workers (POLLARD and McOMIE, 1953) recommend an initial silicone treatment of the glass tube.

Pre-treatment of the column (before separation is attempted) is sometimes recommended. Thus ZECHMEISTER and CHOLNOKY (1950, p. 308) acidify the column with dilute HCl prior to attempting to separate strongly adsorbed cations or with dilute HNO_3 for less strongly adsorbed cations. Usually, full descriptions of the pre-treatment necessary are given along with other procedural data for each method or application.

Choice of adsorbent

The choice of an adsorbent is influenced chiefly by its adsorption capacity, the affinity of the solutes for the adsorbent and the possibility of introducing colouring or fluorescing reagents to aid in the identification of inorganic ions. The ideal is to provide an adsorbent that will permit the desired separation of solutes with a minimum volume of solvent and adsorbent.

Some common adsorbents in increasing order of adsorption are: starch, talc, calcium carbonate, magnesium carbonate, magnesia, activated alumina, activated carbon and activated magnesia. For inorganic ions 8-hydroxyquinoline or other colouring or fluorescing reagents can be used as the "adsorbent". Thorough treatments of this subject are to be found elsewhere (LEDERER and LEDERER, 1957; ZECHMEISTER and CHOLNOKY, 1950; STRAIN, 1942). The method of preparation of the adsorbent will influence its performance. Usually, details of the preparation (or the source) of the materials are stated in the published methods. Adsorbents, prepared to specific standards, are available from most laboratory suppliers.

Application of test solution to column

The solution containing the solutes to be separated is introduced at the top of the column in such a way that the upper layer of the column is not disturbed. A small disc of filter paper, placed on the top of the column, is of great assistance in this regard. The first millilitre is added carefully from a pipette. In elution development, displacement development and the flowing chromatogram, of course, the volume of the initial solution is kept to the absolute minimum. After the solution has been introduced, the solvent is added either in convenient portions to the top of the tube or continuously from a separating funnel or reservoir attached to the column. For frontal analysis, the constant composition solution is added initially with the same care and, later, continuously.

Choice of solvent

A solvent should be non-reactive towards solute or adsorbent and should not be too viscous. Preferably, it should be non-volatile and its eluting power should be such that the separations desired are achieved without the use of large volumes of solvent. It should be readily separable from the solute. For inorganic compounds, non-polar, organic solvents have low eluting power whereas water and ionic solutions (because of their polarity) have high eluting power. STRAIN (1942) gives a list of solvents in order of increasing eluting power which includes: light petroleum, carbon tetrachloride, carbon bisulphide, anhydrous ether, anhydrous acetone, benzene, toluene, alcohols, water, pyridine and organic acids.

Thus for inorganic cations in the series of ZECHMEISTER and CHOLNOKY (1950, p. 308), dilute HCl is necessary to elute As^{3+}, Sb^{3+} and Bi^{3+}, water acidulated with a few drops of dilute nitric acid is required to elute Cr^{3+}, Fe^{3+} and Hg^{2+}. The UO_2^{2+} and the cations below it in the series require only water as the eluant. Thus

As^{3+}, Sb^{3+}, Bi^{3+} } eluted with dilute HCl

Cr^{3+}, Fe^{3+}, Hg^{2+} } eluted with water acidulated with a few drops of HNO_3

UO_2^{2+}, Pb^{2+}, Cu^{2+}, Ag^{+}, Zn^{2+}, Cd^{2+} } eluted by water only

Development of the column

If the column is to be extruded and the zones are to be separated by cutting (after the method of Tswett), development need only proceed until separate zones have been obtained. Usually the front of a zone is well defined but, at the rear, the concentration of the solute decreases more slowly causing "tailing". If the zones are naturally coloured, their separation will be evident. If they are colourless, normal development may have to be interrupted while some reagent that produces coloured or fluorescent compounds with the adsorbed solutes is passed down the column. Thus Schwab and Ghosh and Schwab and Jockers (as quoted by SMITH, 1953, pp. 31–32) achieved a good separation of As^{3+}, Sb^{3+} and Cd^{2+} on an alumina column which was first treated with tartaric acid. The solutes, after application, were washed with water containing tartaric acid and later were developed with H_2S water. From the top of the column the following zones were observed:

orange zone	Sb^{3+}
upper yellow zone	As^{3+}
lower yellow zone	Cd^{2+}

If a desired purity of the zones prevents the use of such a reagent, it might be necessary to "run" a trial using the reagent and to record the position of the zones after a certain volume of effluent has flowed. After completely eluting the column, the operation is repeated except for the use of the reagent. After the same volume of effluent has flowed it may be assumed that the zones will occupy the same positions on the column as observed in the trial. The column can be extruded and cut as desired.

If the column is to be eluted, it is best to test the effluent with reagents specific for the ions involved. The reagents may either be placed in the effluent or used as spot tests on single drops of the effluent on a plate. A wealth of data is available on suitable reagents. The reader is referred to some comprehensive surveys (LEDERER and LEDERER, 1957; FEIGL, 1954; WELCHER and HAHN, 1955; BRITISH DRUG HOUSES, 1958) and to Table XXXII. Two more recent treatments of the subject are recommended (CANNEY and HAWKINS, 1960; WALDI, unknown date).

Identification and determination of zones

Once separated into zones or eluates, the solutes can be determined qualitatively or semi-quantitatively in numerous ways.

The separated material lends itself to radiometric, spectrographic, colorimetric, and other physical methods. On the other hand, it may be determined gravimetrically or volumetrically by normal chemical methods.

Many semi-quantitative to quantitative determinations of inorganic matter have been made by comparing the width and density of a zone with a series of prepared standards made from solutions of known concentrations. For geological field work, this method has much to commend it.

An adsorption method described by DYKYJ and CERNY (1945) provides an example of a semi-quantitative use of a starch column for the determination of magnesium. It was established that as a

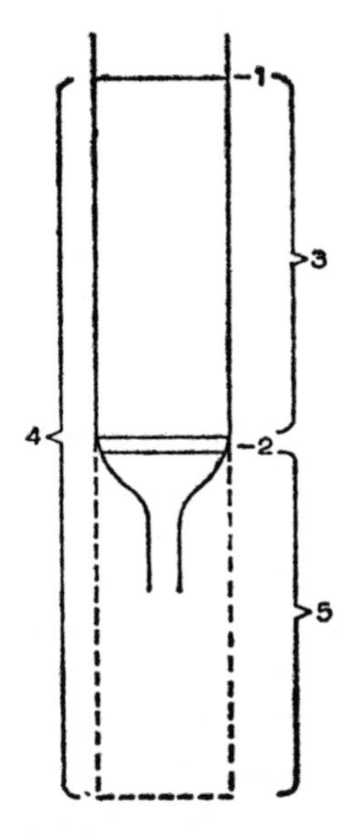

Fig.2. Calculation of R_F value by column data. *1* = point of application of solute. *2* = sintered disc (lower end of column). *3* = volume proportional to the distance travelled by solute. *4* = volume proportional to the distance travelled by the solvent. *5* = retention volume.

magnesium chloride solution of unknown strength, passed down the starch column, a methylene blue ring moved simultaneously. The distance moved by the methylene blue is proportional to the mass of magnesium chloride passing and is independent of the concentration. Later, the subsequent passage of a magnesium chloride solution of known magnesium content provides sufficient data to determine the mass of magnesium in the original solution. Details of this method are given in Chapter 4.

Identification of zones by calculation of R_F values and comparison with published values is of some value where standardisation of the adsorbent, the column and the rate of flow is possible. When this is impossible, R_F values for a number of metal ions in a particular column can be obtained from known solutions. Unknown solutions can be analysed by comparing the R_F values with those obtained with known solutions. The calculations of R_F values on a column require a knowledge of the quantity known as *retention volume* (or retardation volume). This is the volume of liquid passing through the column from the moment a solute is introduced at the top until it (the solute) passes out the bottom of the column. Fig. 2 shows such volumes as proportional to the lengths (since the column diameters are assumed to be equal and uniform). The R_F is the ratio of two of these volumes or distances (see p.28).

PARTITION COLUMNS

Cellulose is the most common support in partition chromatography. It may be obtained as filter paper (chromatography paper) or in tablet or powder form (for example, Whatman cellulose powder for chromatography). In each case, a slurry is prepared and added to the column. When about the desired height of column is obtained, gentle pressure from the top gives the correct degree of compaction. At all times, air should be excluded from the column since bubbles are difficult to remove from the interstices. Silica gel, diatomaceous earth (kieselguhr), starch and rubber are also used as supports. (see Table VI).

TABLE VI

MATERIALS FOR PARTITION CHROMATOGRAPHY COLUMNS

Material	*Description*	*Manufacturer–Supplier*
Cellulose powders		
MN 100	wood pulp	Machery–Nagel und Co., Germany
MN 2100	wood pulp; acid washed	
MN 2100 ff	acid washed; fat and resin free	
MN 2100 WA	acid washed; silicone treated	
MN 2100 Ac	acid washed; acetylated	
MN 2200	linters; acid washed	
MN 300 F 254	fluorescent in U.V. (254 mμ)	
MN 300 GF 254	similar with gypsum binder	
123	cellulose powder	Schleicher und Schüll, Germany
123a	cellulose powder; acid washed	
124	linters powder	
Whatman Ashless	standard or coarse grade cellulose	Reeve, Angel and Co., England
Whatman "B"	standard or coarse grade cellulose	
Silica gels		
Silica gel	finely powdered	E. Merck, A.G., Germany
Silica gel	highly active, suspended in benzene	
Silicic acid SA-1		Bio-Rad Laboratories, Calif., U.S.A.
S.G.	plain	Microchemical Specialties Co., Calif., U.S.A.
S.G. – D5	with 5% $CaSO_4$ binder	
S.G. – D5F	with $CaSO_4$ binder and fluorescence indicator	
Silica gel	plain	Research Specialties Co., Calif., U.S.A.
Silica gel – G	with gypsum binder	
Florisil	synthetic magnesia – silica gel with or without gypsum binder	
Silica gel		J. T. Baker, N.J., U.S.A.
Kieselguhr		
0114	purified; low iron content	Schleicher und Schüll, Germany
Kieselguhr G		E. Merck, A.G., Germany

Normally, it is arranged that the water is the stationary phase and the organic solvent the mobile. However, sometimes it is convenient to reverse this arrangement. Synthetic rubber and diatomaceous earth, treated with silicone, sorb the organic phase and the water becomes the mobile phase. Thus CARRITT (1953) uses granular cellulose acetate as the support for a carbon tetrachloride solution of dithizone as the stationary phase. The mobile phases are the natural waters (containing trace metals) being examined. The metal dithizonates become sorbed on the column and are eluted first by hydrochloric acid and then by ammonia. Greater details are given in the section describing the applications of chromatography to geology (see pp. 67–68).

PAPER CHROMATOGRAPHY

In the thousands of specific separation methods, in which paper chromatography has been used in chemical studies, the great diversity of materials is matched by equally diverse techniques. The general technique of paper chromatography, however, stands out as the simplest and most useful for separations of small quantities of material. Since it can be used in the field, as well as in the laboratory, it seems certain to be very useful in geological studies. For this reason, some emphasis has been placed on paper chromatography in this book, in which it is treated in greater detail than the other forms of chromatography.

The main problems confronting the paper chromatographer in geology are the selection of suitable papers, solvents, identifying reagents and the step by step procedures in the general techniques. Discussion of these matters will be found in the sections that follow. Further details of procedures will be given with each application.

Choice of chromatography paper

With the widespread use of the paper chromatographic method in organic and inorganic chemistry, numerous brands and grades of paper have become available. Surveys of these papers have been

TABLE VII

COMMONLY USED CHROMATOGRAPHY PAPERS

Maker or brand, grade number	*Country of origin*	*Properties*	*Speed or flow rate*
S & S[1] Selecta 2043 b Mgl	Germany	mill finish; medium	medium
W.[2] No. 1	England	medium rough	medium
W.[2] No. 3	England	medium rough	medium
W.[2] No. 4	England	medium rough	medium
Munktell OA	Sweden	rough	slow
Ob	Sweden	rough	slow
D'Arches no. 302	France		
Eaton and Dikeman 7	U.S.A.	medium rough	slow
248	U.S.A.	very rough	fast
613	U.S.A.	very rough	medium
MN[3]214	Germany	medium rough	medium–slow
MN[3]261	Germany	thin rough	medium
MN[3]827	Germany	thick rough	fast
MN[3]263	Germany	thick hardened	slow

[1] S & S = Schleicher und Schüll.
[2] W = Whatman.
[3] MN = Machery–Nagel.

made (LEDERER and LEDERER, 1957; ROCKLAND et al., 1951), and serve as good guides in the selection of a suitable paper for a particular task. Advice from the manufacturers is available and is most helpful. Obviously, the performance of a particular paper will depend upon the solutes to be separated, the solvents to be used and upon the general conditions of use and storage. Impurities in the paper, usually calcium, magnesium, iron and copper, produce some slight interference (irregular "spots" or "patches") with some identifying reagents. Washing with acid is said to produce *purified papers* or *acid-washed papers*. However, it must be remembered that the acid may not be any purer than the paper. Hence, washing may only substitute other impurities for the ones it removes or may release new impurities from the paper fibres. Increased flow rates and decreased strengths are often the legacy of washing. Some wet-strength papers, such as Whatman no. 54 and Macherey–Nagel 263,

TABLE VIII

EXAMPLES OF CHROMATOGRAPHIC SOLVENTS

Paper[1]	*Solvent*	*Use*	*Reference*
S & S 2040 b	(1) methanol + 1% 15 *N* NH_4OH	separation alkali metals	*Anal. Chim. Acta*, 22: 142
	(2) methanol + 5% 15 *N* NH_4OH	separation alkali metals	*Anal. Chim. Acta*, 22: 142
	(3) butanol 5, acetone 2, water 1	inorganic anions	*Z. Anal. Chem.*, 173: 208
W1	phenol 4, water 1 + NH_3 + KCN	R_F of iron chelates	*J. Chromatog.*, 8: 261
W3MM	(1) aq. 1.5 *N* KI 1.5 *N* H_2SO_4	metal iodide complexes	*J. Chromatog.*, 7: 366
	(2) 2 *N* HCl	metal iodide complexes	*J. Chromatog.*, 7: 366
	(3) butanol 2 *N* HCl	metal iodide complexes	*J. Chromatog.*, 7: 366
	(4) ethanol, HCl, water, various proportions	metal iodide complexes	*J. Chromatog.*, 7: 366
W1	(1) methanol: ethanol 1/1	lithium in ores	*Bull. Soc. Franç. Minéral. Crist.*, 80: 181
	(2) ammonia	boron	
	(3) acetone + 20% HCl + 10% water	beryllium	
	(4) acetone 90, H_2O 8.5, HF (40%) 1.5	tantalum	*Bull. Soc. Franç. Minéral Crist.*, 80: 275
	acetone 50, hexone commercial 40, nitric acid (D1.33) 10, water 10	arsenic in ores	*Compt. Rend.*, 253: 1980
W1	(1) butanol 50, 10 *N* HCl 25, 48% HF 1, water 24	separation inorganic ions	*J. Chem. Educ.*, 38: 400
	(2) ethanol 30, methanol 30, 2 *N* HCl 40	separation inorganic ions	*J. Chem. Educ.*, 38: 400
	(3) acetone 90, conc. HCl 5, 48% HF 1, water 4	separation inorganic ions	*J. Chem. Educ.*, 38: 400
	ethanol 60 pyridine 20 water 16 conc. ammonia 4	R_F values inorganic anions	*Anal. Chim. Acta.*, 23: 30

[1] S & S = Schleicher und Schüll, W = Whatman

preserve their strength when wet. Their use under moist tropical conditions seems justified.

The direction of machining of some papers influences their properties. It thus becomes necessary to use the paper so that the direction of flow and direction of machining have a constant mutual relationship. Manufacturers indicate the direction of machining on their product, Whatman by an arrow on the container, Schleicher und Schüll by a water-mark in the paper.

Choice of solvents

A great many solvents have been used mainly for specific separations. A suitable solvent should achieve the desired separation of ions and should produce small round spots without "tails" (or "comets"). For geological work, especially at base camps, solvents, preferably, should be single phase mixtures, should have an effective life of at least a few days and should be stable in both the liquid and vapour phases.

Solvents chosen for inorganic chromatography are commonly mixtures of organic solvents and mineral acids with varying proportions of water. An analysis of some solvents used in recent inorganic separations is recorded in Table VIII.

The choice of solvents often dictates other essential apparatus. For example, if HF is included in the solvent, vessels need to be made of polythene. Since most solvents contain some volatile matter vessels should be securely sealed by lids. Some constituents of solvent mixtures, for example butanol and acetone, are toxic and care should be taken not to inhale the vapours.

Preparation of the chromatograms

Generally, chromatography paper is supplied in sheets about 22″ × 18″ and 22″ × 27″. Half and quarter sizes are easily obtained from these (or any other size, of course) by cutting (not tearing) to suit the size of the solvent vessel being used. Circles of various diameters, reels of various widths and lengths and paper in special shapes are also supplied. Rough vertical edges on the paper accelerate the flow of solvent and ions near that edge. All manner of contamination of the paper, including creasing of the paper with the fingers or resting the hands on the paper, should be avoided.

To permit *ascending development*, narrow strips of paper can be suspended from hooks in the lid of the solvent vessel (see Fig. 3) or rectangles of paper can be formed into gaping cylinders with paper clips and stood in shallow pools of solvent in the vessels. Separations can be carried out with thin strips of paper in test tubes.

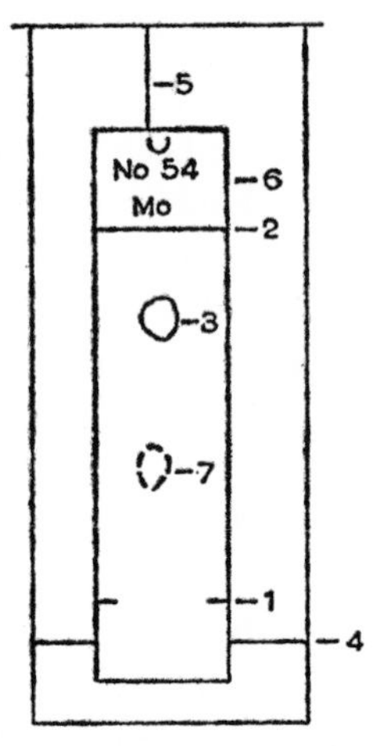

Fig. 3. Paper strip (chromatogram) in development vessel. *1* = point of application of test solution. *2* = solvent front. *3* = spot due to coloured complex. *4* = level of solvent. *5* = hook from lid. *6* = sample data. *7* = position of colourless complex (to be detected later by reagents).

Prior to being placed in the solvent, however, the paper should be marked with a faint broken pencil line drawn parallel to and about 1.5–3 cm from the lower edge. This line indicates the points of application of the test solutions. It is placed so that it rests just above the level of the solvent when the paper is stood or suspended in the solvent. Other necessary data may be written in pencil below the point of application or at the top of the paper.

The test solution is transferred to the paper at the line of application by a micro-pipette (especially for semi-quantitative work). However, for qualitative work a small glass capillary tube can be used equally well. The solution should be applied as a small spot (not greater than 2.5 mm in diameter) or as a thin streak about 1.5 cm long. It is desirable to "run" chromatograms at least in duplicate, taking the mean result or rejecting the chromatogram if the results differ by more than two per cent. For this reason, several fairly uniform applications of test solutions should be applied to the same paper sheet. Too much test solution tends to produce large spots and comets. In a direction at right angles to the direction of flow, most thin papers become overloaded if more than about 0.5–10 mg of solute per centimetre is applied (VAN ERKELENS, 1961). Thicker papers are designed to carry greater quantities of solute. The test solution should be thoroughly dried before being "irrigated" in the solvent. Some authorities (JERMYN and ISHERWOOD, 1949; BLOCK et al., 1958) also recommend that the chromatogram, so prepared, should be suspended in the vapour above the solvent for some hours to equilibrate before the lower edge is immersed in the solvent.

The rectangle of paper permits *parallel chromatograms* to be "run" under almost identical conditions. Thus known and unknown sample solutions are compared by chromatographing them on the same sheet. On a sheet approximately 11″ × 9″, the author often "runs" six or seven parallel chromatograms. In each of these chromatograms three or four metal ions are often included in the test solution or are applied, in turn, to the point of application. The paper should be allowed to dry between each addition, however, to prevent the test solutions spreading. Blasts of warm air from electric dryers can be applied, but the author has found this unnecessary. The moisture content of the paper is disturbed to some extent by the warm air.

Development

In the *descending development* technique, the chromatograms are prepared in a similar way. After equilibration in the chamber, the upper edge of the paper is bent over the edge of a trough containing the solvent. The paper then hangs vertically downwards from the trough. Small glass beads may be hooked on to the lower end of the paper to keep it taut, and a glass plate may be used to hold the paper in the trough. The solvent, after flowing down the paper under the influence of gravity, may be allowed to drip off the end of the paper. Alternatively, absorbent cotton may be used to absorb the effluent, or the lower edge of the paper may be serrated to permit the solvent to drip off. Sometimes the paper is cut so that a constriction of flow occurs at the point of application (LEDERER and LEDERER, 1957, p. 133).

Thus the solvent moves up the paper under capillarity in ascending development and down the paper under gravity in descending development. Flow rates are quicker in the latter. The solvent usually advances up or down the paper with two fronts, *the organic front*, behind which lies a zone of fairly water-free organic solvent, and the *aqueous front*, behind which the paper is saturated with the solvent mixture. From the time development begins it takes some little time for a condition of equilibrium to be reached between solute, solvent and the paper. Hence, the greater the distance flowed by the solvent, the more the separation of ions represents that achieved under conditions of equilibrium. Solvent movements of 10 cm give reproduceable R_F values but movements of 15 cm are more reliable and are favoured by the author. The time of development to achieve these flow distances varies with different solvents and may be from 30 minutes to 18 or 24 hours. Development times of 9–15 hours are inconvenient in laboratories in which work ceases at night. Experience with a solvent enables one to estimate the time of development and normally the vessel should not be uncovered during that time. Laboratory timers and even wrist alarm watches assist in the correct timing of the desired duration of the development.

After development, the chromatogram is removed from the solvent vessel and is placed horizontally on absorbent paper. The solvent front should be quickly marked in pencil. Particularly under field conditions, when volatile solvents (such as acetone) are being used, the solvent front can disappear rapidly before it can be marked. To avoid this, a few smears of a convenient dye (such as rosolic acid in ethanol, 0.1 mg/ml, or even some ball-point pen inks) are placed on the line of application during the preparation of the chromatogram. The dye travels with the solvent front and hence marks its ultimate position (RITCHIE, 1961, p. 401).

Temperature gradients across the solvent vessel tend to produce irregular flow by some solvents. Especially under field conditions, solvent vessels should be protected from draughts and direct sunlight by covering with a larger vessel.

Following the original use in organic chemistry (CONSDEN et al., 1944), *two-dimensional development* has been used in many inorganic separations (KNIGHT, 1959; LACOURT et al., 1949). The method consists of "running" a chromatogram in one direction in the normal way, except that the point of application is near one corner of a rectangle of paper. Later, a second solvent is allowed to flow over the paper in a direction at right angles to the first. The first solvent must either be removed from the paper or be inert towards the second solvent. R_F values can be calculated for the movement in each solvent, but a grid system offers better definition of the position of the spots. The first direction of flow, the *X*-axis, is divided into ten equal intervals which are lettered A–J. The second direction of flow, the *Y*-axis, is similarly divided and numbered 1–10. The position of a spot, after two-dimensional development can be stated in terms of these ordinates (see Fig. 4).

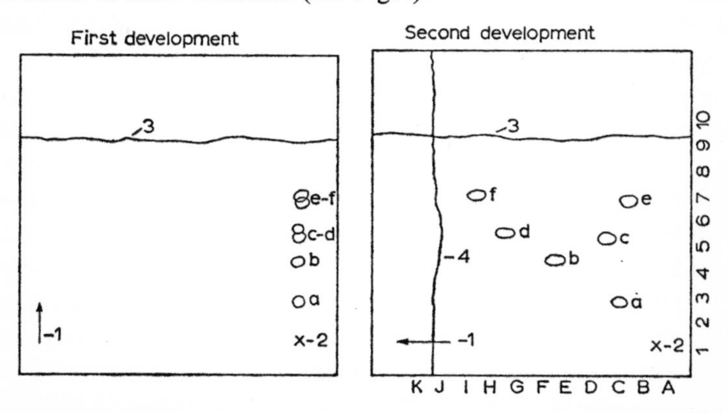

Fig. 4. Two-dimensional paper chromatography. *1* = direction of flow of solvent. *2* = sample solution applied. *3* = solvent front (first solvent). *4* = solvent front (second solvent). *a–f* = spots formed by ions.

Multiple developments, involving the flow of two solvents in the same direction, either in connection with uni-dimensional or two-dimensional development, also effect otherwise difficult separations (ALBERTI and GRASSINI, 1960).

Detection of the ions

The term detection is restricted to the act of locating the positions of the ions on the paper. This is achieved usually by the production of coloured or fluorescent compounds or complexes by the reaction of the ions with suitable reagents. Thus the positions of the ions are indicated by the production of coloured or fluorescent "spots". To some extent, non-fluorescent ions appear as "dark patches" under ultra-violet light.

Many ions form coloured complexes with the solvents without the aid of other reagents. In Table XXVII, ions forming coloured complexes in certain solvents are listed. The spots, sometimes fleeting, should be marked on the chromatograms as soon as possible after they are taken out of the solvent. When transparent glass vessels are used, non-coloured spots of different reflectance from the rest of the paper are seen sometimes on the chromatograms during development. Probably they are caused by a greater concentration of some ions which retain considerable amounts of water thereby creating greater transparency in the paper. These spots fade very quickly when the chromatograms are removed from the vessel. It is possible to mark the position of such spots and the position of the solvent front at that moment, and of the point of application by using a crayon on the outer wall of the glass vessel. The concentration of the ion determines whether or not such spots or coloured spots will be visible on the chromatogram. Hence, this method of identification is not conclusive. The absence of a spot on the chromatogram at this stage does not prove the absence of the ion. Reagents are more positive in this respect especially as the best technique is designed to keep the quantity of ion to a minimum (to reduce the size of the spot). Hence, the data in Table XXVII, obtained from incidental observations made during R_F determinations, may not be complete. Other ions used in greater concentrations may also give coloured spots. Conversely, some of the ions listed may not always produce a coloured spot because of low concentration. Nevertheless, the detection of an ion in this way often saves the use of a reagent.

In analytical chemistry, a great variety of colour forming or fluorescent reagents for inorganic ions has been discovered and recorded (WELCHER, 1947; FEIGL, 1954; BRITISH DRUG HOUSES, 1958; WALDI, unknown date). Of these, some are particularly suitable for paper chromatography in that they can be sprayed on to the chromatogram. Circumstances will indicate whether a specific reagent for a single ion is required or whether a general reagent for a group of ions is needed. Tables XXVIII–XXXII list some general and specific reagents and, in some cases, give brief instructions for their use. More detailed instructions will be found in the references cited.

Calculation of R_F values

Distances to be used in the calculation of R_F values are measured from the point of application to the centre of the spot and to the solvent front. Sometimes, when "comets" are formed, it is difficult to define the centre of a spot. Then it seems best to measure to the top of the combined comet and spot. Large spots, produced by excessively high concentrations of the metal ions in the test solutions, lead to inaccuracy in R_F calculations. Hence, on all occasions, care should be taken to keep the concentrations to the minimum required to produce recognisable "spots".

If ascents or descents of about 15 cm are effected, the probable error of the R_F value (up to 2% normally) does not require the distances travelled by the ions and by the solvent to be measured with any greater accuracy than $\pm$ 0.05 cm.

The probability of small variations in R_F values is to be expected and is explained, to some extent, in terms of variation in concentration of the test solutions, temperature differences, differences in shape and size of the vessels used, variation in paper from batch to batch, direction of machining of the paper and the inherent inhomogeneity of the paper.

Chromatographic profiles

The use of the R_F values of an ion in a number of solvent mixtures in the identification of the ion creates some difficulty since the numbers themselves are not strikingly diagnostic. So, in the problem of R_F representation, R_F diagrams or "spectra" (REIO, 1958) have been introduced into organic chemistry. The present author, when introducing this concept into metal ion chemistry, made a case against the use of the term "spectrum" and proposed the term "chromatographic profile" (RITCHIE, 1961). The chromatographic profile of an ion is obtained by plotting and joining its R_F values in the solvents on adjacent columns. If the solvents are well chosen, the chromatographic profile of an ion is characteristic and permits identification of the ion (see Fig. 5).

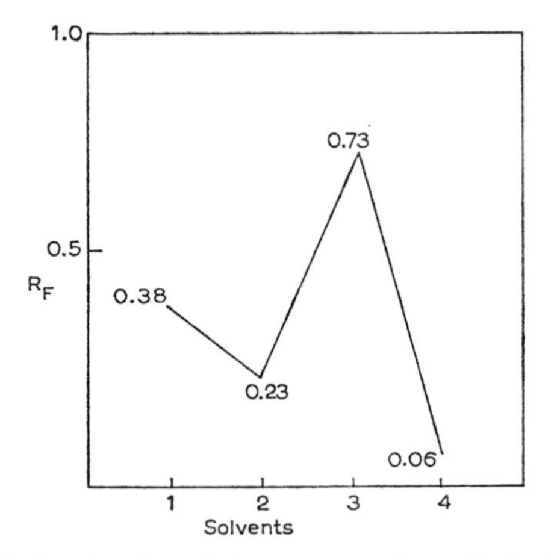

Fig.5. Construction of chromatographic profile for Cr^{3+}.

Identification of Ions

In paper chromatography, identification of an unknown ion rests mainly on its R_F in a solvent or its chromatographic profile derived from R_F values in a number of solvents. Confirmation is obtained, however, from the colour of the spot on the chromatogram and the colour or fluorescence after treatment with one or more reagents.

When several metal ions are present in the one test solution, the chromatograms from all the solvents should be examined at the same time. In this way, the matching characteristics of each particular ion (or spot) can be observed. Such matching characteristics include:

colour of the spot as the chromatogram is lifted out of the solvent,
time for the test reaction to proceed,
colour changes as the reactions proceed,
slight difference in the colour of the spots,
degree and hue of the fluorescence,
effect of ammonia spray on the fluorescence.

Table XXXIV "will be found of assistance in that ions of approximately the same R_F values in any one solvent can be seen at a glance. The R_F value in the next solvent usually eliminates all but one or two of the possibilities of the first solvent and so on" (RITCHIE, 1961).

Parallel chromatograms of known and unknown ions provide a sound method of identification especially under unusual conditions. "Shortcut" methods can be adopted when the presence (or absence) of one or more ions in a test solution is required to be demonstrated. Since each solvent has some advantage of separating certain ions from one another, a careful selection of the solvent and then a few known solutions "run" parallel to the unknown, is all that is necessary to confirm the presence (or absence) of a metal.

ION EXCHANGE

In the field of chromatography, the materials available are nowhere so standardized and so specific in respect of properties and performance as in ion exchange (see Table IX). Hence a number of delicate and refined applications to geology have been made. A precise technique is necessary to achieve success in these methods.

Ion exchange columns

Ion exchange columns offer fewer problems of packing and operation than adsorption and partition columns. The uniformity of the bead size limits and the generally larger size of the particles

TABLE IX

ION EXCHANGE RESINS

Strongly acidic cation exchangers	*Weakly acidic cation exchangers*	*Strongly basic anion exchangers*	*Weakly basic anion exchangers*
Dowex 50		Dowex 1 and 2	Dowex 3
Dowex 50 W		Dowex 21K	
Zeo-Karb 315	Zeo-Karb 216	De-Acidite FF	De-Acidite E
Zeo-Karb 215	Zeo-Karb 226		
Zeo-Karb 225			
	Amberlite IRC.50	Amberlite IRA-400	Amberlite IR-4B
		Amberlite IRA-410	Amberlite IR-45
Amberlite IR-120			
Whatman[1] P 10	Whatman CM 30	Whatman DE 50	Whatman EA 50
Whatman P 40	Whatman CM 70		Whatman EA 30
Whatman P 70			

[1] Whatman exchangers are all "modified cellulose ion-exchangers".

(beads), compared to that of adsorbents and partition column supports, make the packing technique very simple. The resulting columns are very uniform.

Resins are supplied in a moist condition and should be maintained in that condition while stored. Some resins (for example, Dowex) are supplied ready for use while others (for example, Permutit) need to be *regenerated* before use. Thus the term "regeneration" sometimes refers to the initial preparation of the resin *before use* and universally to the restorative treatment of the resin *after use* and before its renewed use. Resins that have been allowed to dry often contain broken and split beads. These fines should be removed by repeated washing with water and decanting while the fines are still in suspension. For all treatment with water, either deionized or distilled water must be used.

The ratio of height of resin column to diameter should not be less than ten to one. One manufacturer recommends a ratio of thirty to one for a one inch diameter column. Since a necessary step in the preparation is "back-washing", which produces a fifty per cent expanded height, the glass column should be about five feet long.

Burette type glass columns are the most suitable for the kind of operations that lend themselves to geological applications. Polythene and plastic tubing can be used in field work. Burettes, normally, do not require calibration for uniformity of cross-sectional area. An extension tube should be added to the top of the burette to cope with and to dispose of the water used during back-washing.

The resin needs a supporting pad to prevent it from being washed out of the column. A pad of glass wool is often sufficient. Small glass beads (about 3 mm in diameter) may be added on top of the glass wool to use up the space below the graduations on the burette. Then the length of the resin column can be read directly from the graduations on the burette.

Before adding the resin to the burette a few cubic centimetres of water should be added. Then the resin slurry is introduced into the column, being washed in with more water. The resin is then back-washed for several minutes with water which is made to flow upwards at such a rate that the resin beads assume a turbulent motion and undergo a 50% volume increase. Back-washing should always be started very slowly in order to prevent the glass-wool support being forced up the tube. After back-washing, the resin is allowed to settle, whereupon the water in the column is drained down to within one inch of the top of the resin. An alternation of back-washing and draining is continued until resin height, after several drainings, is constant. Once packed in this way, the resin should always be kept covered by liquid.

To *regenerate* the resin one should follow the instructions issued by the suppliers. Concentrated solutions of regenerants are used since they are more efficient and require smaller volumes which, in turn, reduce the time for regeneration. Sometimes a countercurrent regeneration is used, the direction of flow of the regenerant being the opposite of that of the solute(s) in solution. This operation ensures that fewer unwanted ions remain on the resin. The rate of the countercurrent flow should be slow enough to avoid undue expansion of the resin.

If many separations are envisaged, a system containing a supply tank, a regenerant tank, a back-wash tank and an overflow tank, with connecting tubes and clamps, is most advantageous. A constant head tank permits a more regulated flow of solvent.

Multiple columns, through which the sample solution is passed in turn, are more effective than a single column especially if the columns are made progressively smaller in diameter.

Concentrations and bulk of the sample solution should be kept low (if this is possible). The total ion exchange capacity of the resin should be taken into consideration. The sample solution should not contain solutes equivalent to more than one third to one half of the total exchange capacity of the column. Hydrogen ion concentration should be controlled, pH less than three reducing the effective capacity of the column considerably. The recommendations of the manufacturers regarding pH should be followed. For "Zeo-Karb 226", the capacity (in the presence of 0.1 *N* NaCl) increases from 0.8 mequiv./g(dry) at pH 5 to 9.0 mequiv./g(dry) at pH 9 (PERMUTIT Co., unknown date). Care should be taken to avoid the introduction of solutions which also contain suspended matter or which may lead to the precipitation of matter on the resin.

The flow rate, measured as the effluent from the column, will depend on the dimensions of the column, the pressure produced by the head, the suction at the lower end and the opening provided by the clamp or tap. Fast rates reduce the efficiency of the resin while slow rates tend to give the best separations, but, of course, lengthen the duration of the process. A normal flow rate is one that causes an advancing front in the column to move at five centimetres per hour. Expressed another way, this means a normal rate of flow of effluent from a column of one inch diameter is about ten to fifteen millilitres per minute. From a one square centimetre column, the rate is about one or two millilitres per minute. Rates of flow are generally stated in the descriptions of specific methods. In original work, they must be determined by experiment.

Ion exchange resins are stable over a fairly wide temperature range, most being stable from 5°C–50°C and some up to 150°C. Normally, hot solutions should not be used. Otherwise, resins can be used under normal laboratory and field conditions.

The careful selection of a resin (see Table IX), the packing of the column and the attention to the matters already discussed are preliminaries to the actual separation.

The sample solution can be allowed to flow down the column in such a way that those ions so desired will be sorbed on to and into the resin. Finally a few millilitres of water should be added to carry the last of the solution into the resin column. As the solution moves down the column, the solute ions are exchanged for exchangeable ions on the resin. The upper region of the column becomes saturated with solute ions while lower regions still have available exchangeable ions. When two or more different solute ions are present, they will be sorbed at different rates and, under favourable conditions, separation of the different solute ions will be achieved.

The order of selectivity for the ions is stated in data sheets and specifications published by the manufacturers. Thus for "Zeo-Karb 215" (PERMUTIT Co., unknown date, p. 7), the order is:

$$Hg^{2+} < Zr^{4+} < Li^{+} < H^{+} < Na^{+} < K^{+} = NH^{+}{}_{4}$$

and so on. For Dowex 50 (DOW CHEMICAL COMPANY, unknown date, p. 71) the order is stated:

$$Ag > Ca > Rb > K > NH_4 > Na > H > Li.$$

The process of removing the sorbed ions from the resin is called *displacement*, *elution* or *regeneration*. The latter term also implies that the resin is reconverted to its proper ionic form for re-use. Regenerant liquids are acids (for acidic cation exchangers), alkalis (for basic anion exchangers), salt solutions (for Na-form resins) and complexing agents such as citric acid and tartaric acid. The concentration of regenerant solutions is in the range 5–10% for strongly acidic and strongly basic resins and 1–5% for weakly acidic and basic resins.

The displacement type of elution can be used often to desorb selected ions in turn. A low concentration eluant is used first to desorb the ions with the lowest affinity. A second eluant of stronger concentration desorbs the next kind of ion and so on.

When specific applications have been described (ELLINGTON and STANLEY, 1955), information on the eluants will have been stated.

General information is made available in data booklets by the manufacturers.

A common technique is to collect the effluent in a large number of small equal fractions. Automatic and manually operated rotating devices have been designed for this purpose. The effluent may be subjected also to a continuous conductivity test by leading the effluent past two electrodes connected to a galvanometer circuit. If radioactive materials are used, the effluent may be passed through a rate counter.

In ion exchange batch operations, the whole of the solution, containing the solutes to be separated, is mixed closely with a mass of the ion exchanger. After a sufficient time has elapsed, the exchanger is separated by filtration, decantation or other means. This single step operation can be likened to a single step in separation by adsorption or by partition. If the exchanger has a relatively weak affinity for the solutes little separation will be achieved. If, however, there is a strong affinity then a satisfactory separation will result. In the former case, just as repeated partitions are necessary to achieve some satisfactory degree of separation, so two or more steps in the ion exchange operation achieve a better separation by ion exchange. In the first step a "new solution" is treated with an exchanger that has been used once already for the same kind of separation. The high concentrations in the "new solution" achieve a maximum of ion exchange on the once-used resin. In the second step the once-used solution is treated with freshly-regenerated resin. Now the higher exchange capacity of the resin favours maximum exchange with the more dilute, once-used solution.

Batch operations are simple to perform but are not as efficient as column operations.

Ion exchange papers

A rather limited ion exchange capacity for ordinary chromatography paper has been claimed (PICKERING, 1959). Some attempts have been made to increase the exchange capacity by chemical treatment. The impregnation of ordinary chromatography paper by dipping it in colloidal suspensions of ion exchange resins followed

(M. LEDERER, 1955; HALE, 1955). These early forms of ion exchange paper gave indications that the ion exchangers acted in much the same way as the corresponding resins in columns.

At least two manufacturers (Schleicher und Schüll and Balston [Whatman]) now supply ion exchange papers. The former supplies cation exchangers CAM and P, and anion exchangers DEAE and ECTEOLA. The latter supplies:

Cellulose phosphate, a strongly acidic cation exchanger.

Carboxymethyl-cellulose, a weakly acidic exchanger.

Aminoethyl-cellulose, a weakly basic anion exchanger.

Diethylaminoethyl-cellulose (DEAE), a strongly basic anion exchanger.

Ecteola cellulose, a weakly basic anion exchanger.

Ascending and descending developments have been used with these papers. Two-dimensional separations of inorganic ions have been reported (KNIGHT, 1959). It seems that ion exchange paper techniques combine the partition and ion exchange processes with the great versatility of paper chromatography. It is possible to exploit this versatility, for example, by achieving two-dimensional separations using partition processes with organic solvents in one direction and ion exchange processes with aqueous solutions in the other. Otherwise, the techniques for the development and detection of ordinary paper chromatography are the same for these newer papers. The higher cost of the new papers will require that they be used with strict economy and on worth-while projects.

Modified cellulose exchangers in the form of floc provide for their use in both batch and column operations. In the powder form, they are suitable for column operation.

THIN LAYER CHROMATOGRAPHY

In recent years, the "chromatoplate" process, which was introduced in 1951 (KIRCHNER et al., 1951), has been made more popular and practical by STAHL et al. (1956), who have simplified and standardized the operating techniques. Onto glass plates, adsorbents and partition supports are spread uniformly in thin, adherent layers which may be adjusted in thickness from 50–3000 μ. Hence the term "thin layer chromatography" (T.L.C.) has become widely used.

The method has several advantages over column and paper chromatography. The time for development is very short (commonly from five minutes to one hour). The glass support is resistant to most corrosive reagents and detection by fluorescence is more definite in the absence of the vague fluorescence of the chromatography paper itself.

Thin layers have been prepared from silica gel, alumina, kieselguhr, cellulose, DEAE–cellulose (ion exchange), carbon and polyamide powder. Generally, the material should be very fine (minus 200 mesh) and well graded (by sedimentation, if necessary). Some of the materials can be activated by heat (at about 110°C), such activation being done best on the newly spread layers during drying. Activation decreases steadily after the first few hours. Binding and fluorescing substances can be added to the adsorbent.

The uniform thin layer is achieved by applying the adsorbent in the form of a slurry from a reservoir tank which is incorporated in a "spreader". Spreaders are available as 250 μ fixed spreaders or as variable-thickness spreaders (50–3000 μ).

Techniques for both ascending (STAHL et al., 1956; PETSCHIK and STEGER, 1962) and descending development (MISTRYUKOV, 1962) have been used. The detection techniques for paper chromatography are equally suitable for thin layer chromatography.

Most of the early developmental work with thin layer chromatography involved organic separations. The author has not encountered any applications of thin layer chromatography to geology. The simplicity and adaptability of the method point to developments sooner or later in this direction.

THE CHROMATOBOX

A simple piece of apparatus, which has been introduced recently, is the "chromatobox" of Barrollier of Schering A.G., West Berlin. It consists of a polythene box (7.5 cm × 7.5 cm × 4.5 cm) which contains a small compartment to accommodate from 5–8 ml of solvent. This compartment leads over a weir to a larger compartment which accommodates a strip of chromatography paper (40 cm × 6 cm) which is rolled with a band of Teflon of similar length. The surface of the Teflon is raised thus separating the layers of the paper from each other and also permitting uniform flow of the solvent along the paper. The roll is held secure by an open polythene ring. The paper is rolled so that one end of it protrudes to such an extent that it can be led over the weir into the solvent. Here it is held between two small glass plates, which, in turn, are held together by a glass rod pressing against them. A tightly-fitting polythene lid is provided. The test solution is applied to the paper in the usual way on the horizontal surface near the weir. After the sample dries, the solvent is added and development is carried out in the closed chromatobox. Methods of detection and identification of ions on the chromatogram are the same as in ordinary paper chromatography.

In preliminary tests, the present author has obtained excellent separations of the common metals from a series of prepared mixed solutions of their salts. For both laboratory and field use, the chromatobox seems to provide a very simple and efficient method of paper chromatography.

CHAPTER 4

Applications of Chromatography to Geology

Since the borders of chemistry, geochemistry and geology overlap, no attempt will be made to assign any priority to applications of chromatography to geology. To do so would start arguments that would be bound to rest upon definitions of the vague boundaries between these subjects. The author prefers to present the applications in such groupings and in such order that limit repetition.

Natural waters, being weak solutions of many metal ions, lend themselves to chromatographic column analysis for, by this method, the solute ions become concentrated on the stationary phases during the passage of large volumes of water. Chromatographic methods of water sampling and stream analysis are now well established in the field of hydrogeochemistry. Other fields of geologic enquiry to which chromatography has been applied include:

Soil, rock, ore and coal ash analysis.

Separation and determination of like or interfering metals.

Determination of metals in ores.

The determination of the metal content of minerals as a basis of mineral identification.

The determination of metals in plants or plant ash.

One comes to recognise, too, that in mineral, ore and rock formation, the processes which we have termed chromatographic processes play important roles.

NATURAL WATER SAMPLING AND STREAM ANALYSIS HYDROGEOCHEMISTRY

CARRITT (1953) has shown the importance of data on trace metals in natural waters in understanding the biological and catalytic processes that operate therein. The dearth of information generally and the problems of analysis of natural waters containing metals in widely different concentrations are outlined briefly. Carritt points out that 94.2% by weight of dissolved solids of sea water is made up of six elements (chlorine, sodium, magnesium, sulphur, calcium and potassium). Forty-three elements have not been detected in sea water. The remainder, however, may be grouped as follows:

in concentrations	$10^{-4} - 10^{-6}\%$	13
in concentrations	$10^{-6} - 10^{-8}\%$	17
in concentrations	$< 10^{-8}\%$	8

For modern analytical methods to be applied, with requisite precision and accuracy, the natural waters need to be concentrated by factors up to 10,000. The method of concentration must not nullify the precision of the analytical method.

Carritt devised a partition column in which a solution of dithizone in carbon tetrachloride formed the stationary phase which was held in a column composed of cellulose acetate. The natural waters, flowing through the column, act as the mobile phase. Lead, zinc, manganese, cadmium, cobalt and copper were completely removed from both natural waters and prepared solutions of known solute concentration.

The column is prepared from cellulose acetate (Fisher Scientific Co., no. C-215) which is crumbled and sieved. Two fractions give satisfactory results, namely:

(a) minus no. 18 mesh (1,000 μ opening) and plus no. 25 mesh (750 μ opening),

(b) minus no. 25 mesh and plus no. 35 mesh (500 μ).

10 g of the sized cellulose acetate are treated with 100 ml of a carbon tetrachloride solution of dithizone (0.5 g/litre). The slurry is heated (with constant stirring in a 500 ml beaker on a hot plate) until, due to loss of carbon tetrachloride by evaporation, the bulk of the material passes from a deep blackish green to a light green. At this marked transition the material is dry enough to be poured and packed into the column. Above a glass wool plug, the column is packed first with approximately 2 g of untreated cellulose acetate

which is tamped firmly. Then 3 g of the treated cellulose acetate are packed firmly and the top of the column is finally plugged with glass wool. Carbon tetrachloride (3 ml) is added to the top of the column and allowed to percolate downwards under the influence of a vacuum collecting bottle. This wetting of the column is reported to be most essential. Wash the column with *M* HCl (100 ml) followed by demineralized water (250 ml).

The pH of the natural water sample is adjusted to 7.0 ± 0.1. Alkalinity of the sample seriously effects the recovery of manganese (which should be in the divalent state). The sample is drawn through the column by a vacuum sufficient to give a flow of 2 litres per hour. Rates up to 6 litres per hour can be used. A large collecting bottle is needed to nullify the pressure fluctuations caused by the vacuum pump. A colour change moving down the column indicates when the column has reached saturation. Sample volumes of up to 10 litres may be treated on one column.

Elution from the column is effected by *M* HCl (50 ml) which completely removes Pb^{2+}, Zn^{2+}, Cd^{2+} and Mn^{2+}. Then 15 *N* NH_4OH (50 ml) removes Cu^{2+} and Co^{2+}. After this last operation the column cannot be re-used. Concentration factors of up to one thousand have been achieved by this method.

Carritt determined Mn^{2+} spectro-photometrically and the other metals polarographically. Recovery tests, made by passing prepared solutions of known composition, indicated recoveries of from 93–114%. Simplicity, efficiency and the little need for attention are claimed as advantages that this method has over liquid–liquid techniques.

COULOMB and GOLDSTEIN (1956a), in their researches into geochemical prospecting for uranium, report a method that uses an ion-exchange technique to collect the uranium from natural waters. The procedure is as follows.

Duplicate columns (8 cm long $\times$ 0.8 cm diameter) are packed with Amberlite IRA 400 or IRA 410 and stoppered with filter paper. The water samples (250 or 500 ml) are treated with 1 or 2 ml (respectively) of saturated solution of sodium bisulphate. After filtration, the samples are passed through the columns slowly (1 drop per second).

For waters with low concentration 1 or 2 litres of sample may be treated. The resin needs to be prepared as follows prior to use:

(1) wash with alcohol,
(2) rinse with water,
(3) stir in polythene beaker with 3 *N* NaOH,
(4) wash with water until free of alkali,
(5) stir with 3 *N* HCl,
(6) wash until free of acid,
(7) stir with 3 *N* NaOH and soak 12 hours,
(8) wash until free of alkali.

After the uranium has been concentrated onto the resin, the latter is transferred to a small silica capsule and heated until it begins to go black (too rapid heating causes decrepitation). Concentrated nitric acid (2 ml) is added and the whole is warmed until the resin is decomposed and carbonized. Heating is done from above by infra-red lamps but, for the final cindering, heat is applied from beneath the capsule. Finally 2 *N* HNO_3 (0.5 ml) is added to dissolve the residue. Determination of the uranium content is made fluorometrically.

If 500 ml of water is taken as the sample, the method is capable of determining uranium concentrations down to 0.1 microgram[1] per litre in the original sample.

COULOMB and GOLDSTEIN (1956a), following their researches into hydro-geochemical prospecting for uranium, report on a number of tested field methods that have "proved sensitive, selective, simple, rapid and economical".

The uranium content of rock samples is taken into solution by treating the weighed pulverized rock with 2 *N* HNO_3.

The final fluorimetric determination of uranium requires a total absence of fluorescence inhibitors such as Fe^{3+}, Cr^{3+}, Mn^{2+}, Co^{2+}, Ni^{2+}. Therefore the uranium is first extracted from the sample solution by ascending paper chromatography on Whatman no. 1 in the form C.R.L. (cut to give 10 small paper strips 1.5 cm wide, 10 cm long separated by gaps in the paper). An exact volume (0.05

[1] $1\ \gamma = 1$ microgram $= 1\ \mu g = 1 \cdot 10^{-6}$ gram.

ml) of sample solution is then applied to the paper and the chromatogram is developed with the solvent which is tributylphosphate and white spirit (mineral turpentine) in equal quantities stirred with 2 N HNO_3 (quantity not specified). Finally, the uranium spots are cut from the chromatograms and determined fluorometrically.

COULOMB (1957) further reports on the method in the light of field experience in testing natural waters for uranium in Guiana. The uranium content of the waters was fixed on small resin columns, which were easily transported to a mobile laboratory. COULOMB and GOLDSTEIN (1956b) report similarly.

WARD et. al. (1960) used a chelating resin A-1 to collect molybdenum from a natural brine solution. Elution with dilute alkali permitted the molybdenum to be determined colorimetrically by reaction with thiocyanate. Saline material and plants were analysed similarly. Aqueous solutions of the salts and dilute acid solutions of the ashed plants were passed down the column and their molybdenum content was determined in the same way. In the early stages of development of the method some difficulty was experienced with some variation in the quality of the resins from batch to batch. This has now been overcome (F. N. Ward, personal communication, 1962).

A field method developed by WARD and MARRANZINO (1957) for determining traces of uranium in natural waters uses a phosphate precipitation to collect the uranium from the bulky water samples onto filter paper pads. At a base-camp or laboratory, the sample on the pads is dissolved in a nitric acid-aluminium nitrate solution and an aliquot is chromatographed on paper by the ascending technique. Estimation of the uranium content is made by comparison with prepared standard chromatograms.

Standard reagents and standard solutions are prepared as follows: Phosphate reagent: sodium phosphate $Na_2HPO_4.12H_2O$, a saturated aqueous solution. E.D.T.A.: 5% aqueous solution of ethylenediaminetetraacetic acid, disodium salt (FLASCHKA, 1959). Nitric acid-aluminium nitrate reagent: dissolve 80 g aluminium nitrate ($Al(NO_3)_3.9H_2O$) in a mixture of 25 ml HNO_3 (s.g. 1.42) and 75 ml water. Standard uranium solution 0.1%: Dissolve 0.211 g uranyl nitrate hexahydrate in 100 ml nitric acid–aluminium nitrate reagent. Prepare dilute standard solutions by adding 0.1, 0.2, 0.4, 0.8, 1.5, 3.0 and 6.0 ml of the standard solution (0.1%) to separate 10 ml volumetric flasks. In each case, the 10 ml volume is made up with the nitric acid–aluminium nitrate reagent. A 0.01 ml aliquot from each of these flasks contains 0.1, 0.2, 0.4, 0.8, 1.5, 3.0 and 6.0 μg of uranium respectively. Solvent mixture: to a 600 ml beaker are added in order: 30 ml ethyl acetate; 6 ml nitric acid (s.g. 1.42) and 0.5 ml water. Stir and cover the clear solution obtained. The mixture is effective for four hours.

Carry out the following method: to a 500 ml sample of natural water add 2 ml of nitric acid (s.g. 1.42). Adjust pH to 2.5 by adding ammonium hydroxide (conc.) drop by drop accompanied by stirring. Long range pH test paper (pH 1–6) is used as indicator. The phosphate reagent is added drop by drop with stirring until the pH lies between 5.9 and 6.1 (as determined with pH test paper 5.8–6.2 range).

If a visible precipitate forms, due to the presence of iron, calcium and aluminium, redissolve it by the addition drop by drop of E.D.T.A. After precipitate is redissolved, adjust pH to 5.9–6.1 range by addition of drops of phosphate reagent or E.D.T.A., (ethylene-diaminetetraacetic acid–disodium salt). Stream waters do not usually form a precipitate and hence this step is not always necessary.

The treated water is filtered through a pad of paper pulp (prepared as a slurry from one Whatman ashless tablet in water and collected in a column on a filter paper circle supported on a perforated disc). After filtration, the pulp and its adsorbed uraniferous contents are placed in a porcelain crucible with 4 or 5 drops of nitric acid (s.g. 1.42). Dry gently on a hot plate and then ash the contents over an open flame and cool. Add 0.5 ml nitric acid, heat gently to dryness and ignite to expel nitric oxide fumes. Cool and repeat the nitric acid treatment, stirring any incrustation from the sides of the crucible into the acid using a small glass stirring rod. Evaporate to dryness and cool. Dissolve residue completely in 0.2 ml of nitric acid–aluminium nitrate reagent (to form the test solution) warming for 20 seconds and avoiding boiling.

Prepare chromatogram on Whatman CRL-1 slotted paper using 0.02 ml aliquots of the test solutions. Dry well for at least 30 minutes in a desiccator. Chromatograph with the prepared solvent mixture in a closed vessel (600 ml beaker covered by the petri dish) for about 1 hour (until solvent front is about 1 inch from the top of the paper). Dry and then spray on both sides with potassium ferricyanide (5% solution). Allow to dry and compare the spots obtained with those observed on chromatograms produced by using standard uranium solutions.

TABLE X

URANIUM RECOVERY FROM PREPARED SAMPLES[1]

Laboratory sample no.	*Uranium added to 500 ml sample (micrograms)*	*Uranium estimated (micrograms)*
CR–2–566	2	1
	5	5
	10	7
CR–2–568	10	10
CR–2–5618	2	1
	4	3
	8	6
	12	11
	32	27
	40	39
	50	49

[1] After WARD and MARRANZINO (1957). Reproduced with the permission of the Director, U.S. Geological Survey.

Each 0.02 ml aliquot contains one-tenth of the uranium content of the original 500 ml water sample. The microgram content of the comparable standard chromatogram is therefore multiplied by 20 to give the uranium content of the water sample in micrograms per litre. The suitability of the method has been demonstrated by laboratory tests. Tables X, XI and XII provide ample justification for this claim.

Ion exchange resin IR-120 was used by CANNEY and HAWKINS

TABLE XI

REPEATABILITY OF URANIUM DETERMINATIONS[1]

Laboratory sample no.	*Uranium p.p.b. (1 American billion = 1,000,000,000)* 1	2	3	4	5	mean
CR-2-566	6	5	4	6	5	5.2
CR-2-5610	16	16	10	12	16	14.0
CR-2-5611	30	30	25	30	30	29.0
CR-2-5612	160	120	160	160	140	148.0
CR-2-568	300	300	300	320	320	308.0

[1] After WARD and MARRANZINO (1957). Reproduced with the permission of the Director, U.S. Geological Survey.

(1960) in hydrogeochemical prospecting for lead, copper, zinc, cobalt and nickel in surface waters of Maine, U.S.A.

The resin (minus 20 mesh–plus 50 mesh) is packed into plastic tubing 15 cm long with an internal diameter of 0.72 cm (25 ml

TABLE XII

COMPARATIVE RESULTS. FLUORIMETRIC AND CHROMATOGRAPHIC ESTIMATIONS[1]

	Uranium p.p.b.	
Laboratory sample no.	*Fluorimetric in laboratory*	*Chromatographic in field*
CR-2-562	0.5	2
CR-2-563	1.5	2
CR-2-564	2.6	2
CR-2-565	2.8	2
CR-2-566	5.9	6
CR-2-567	15	10
CR-2-568	27	25
CR-2-569	60 and 85	60 and 75

[1] After WARD and MARRANZINO (1957). Reproduced with the permission of the Director, U.S. Geological Survey.

volume of resin). Plugs of glass wool are used to confine the resin which is introduced as a slurry. Pre-treatment of the resin includes alternately soaking the resin in 6 *N* HCl and washing it free with demineralized water until free of zinc and iron. (The present author suggests spot tests with dithizone).

One-gallon samples of natural waters are adjusted to pH 7 $\pm$ 0.5 by the addition of hydrochloric acid or ammonium hydroxide, pH test paper being used (it is assumed). The sample vessel is connected to the column and a flow rate of 100 ml per minute is adopted.

Elution is effected by an upward flow of the eluting agent (that is, the reverse direction to flow during collection) at a rate of 1 ml per minute. The eluant used is 20 ml 2 *N* HCl followed by 80 ml of demineralized water. Care should be taken to prevent the introduction of air bubbles into the column. An enrichment factor of 38 is obtained. Colorimetric methods are suggested by the authors for the estimation of the metals in the field.

SOIL, ROCK AND ORE ANALYSIS

Numerous applications of chromatographic methods to soil, rock and ore analysis have been made. It is in this field of geology, where samples are often obtained in extremely remote localities, that the need for reliable field methods is so acute. To a considerable extent chromatography has met this need.

Agrinier has reported specific methods for the qualitative and/or semi-quantitative determination of such elements as beryllium, molybdenum, arsenic, bismuth, selenium, lithium, boron, silver, nickel, cobalt, copper, niobium, tantalum, titanium, uranium, thorium and arsenic in soil, rocks or ores.

Beryllium in rocks and soils (AGRINIER, 1960)

The sample is ground to 75 μ or finer and 1 g is placed in a platinum crucible with an acid mixture (HF [conc.] 4 ml; H_2SO_4 [conc.] 2 ml). Evaporate under an infra-red lamp until no further acid fumes are evolved. The dry residue is dissolved in 4 ml of nitric acid (s.g. 1.33) to which has been added 0.5 ml hydrogen peroxide (30%), the mixture being allowed to stand for 15 minutes. Dry by gentle heating on a sand bath and, finally, by an infra-red lamp. Redissolve in 4 *M* nitric acid and dry rapidly. Cool for some minutes and dissolve the residue in 2.5 ml nitric acid (15% aqueous solution). Stand 5 minutes and then stir well mechanically. Filter on a micro-funnel (2 cm diameter and a pad 2 cm thick). The filtrate is received into a 5 ml polythene beaker. A sample of 0.025 ml of the filtrate is used for the chromatogram. A larger version of Whatman CRL-1 paper with paper strips 3 cm wide and 16 cm high is used. After the sample is applied to the paper and allowed to dry, the chromatogram is suspended in a glass vessel (35 cm × 15 cm × 20 cm high) and allowed to dip into 200 ml of solvent mixture (acetone 130 ml, nitric acid 70 ml). The paper is supported by hooks fitted to the perspex lid which rests on a sponge rubber seal. Development is timed to allow a 10 cm ascent which takes about 1 hour. The solvent only lasts one day. After development, the chromatogram is dried in an oven at 35°C. It is then exposed to ammonia vapour for 2 minutes to neutralize the nitric acid from the solvent. Spray both sides with 2% quinalizarin solution (dissolve 0.2 g of quinalizarin in 10 ml pyridine, make up to 100 ml with acetone and filter). The chromatogram is then exposed to acetic acid fumes for 1 minute which removes the coloration except where a metal lake forms. The beryllium, having an R_F of about 0.45, forms a violet-mauve band above a strong violet band formed by iron and aluminium at about R_F 0.25.

Standard chromatograms are prepared from a solution of $Be(NO_3)_2$ containing 0.1% beryllium. Place 0.0694 g of spectrographically pure BeO in a platinum crucible with 2 ml hydrofluoric acid (25%). 1 ml of HF (conc.) is added. Warm until all oxide is dissolved and dry on a sandbath. The residue is taken up twice in 3 ml nitric acid (conc.) and dried each time. Upon cooling take up in 25 ml nitric acid (15%). Starting with this solution (0.1% Be) the following dilute solutions are made: 0.020, 0.100, 0.0075, 0.0050, 0.0025, 0.0010, 0.0005 and 0.0002%. To simulate the normal solutions obtained from rock or soil samples, 1 g samples of granite, almost free from Be (less than 1 p.p.m.), are dissolved in the manner

TABLE XIII

COMPARATIVE BERYLLIUM DETERMINATIONS (IN P.P.M.)[1]

Prepared samples of known content	*Determination*	
	Chromatographic	*Spectrographic*
0 (Be-free granite)	0	0
9	10	11
22	25	26
30	25	26
45	50	53
90	100	87
120	100–125	90
225	200	179

[1] Reproduced with the kind consent of the author. After AGRINIER (1960).

indicated above, the final residues being taken up in 2.5 ml volumes of each of the dilute standard solutions. Each final sample is filtered and stored in a 0.5 ml polythene bottle. Then 0.025 ml of each final sample is chromatographed and detected as above. Thus a

TABLE XIV

BERYLLIUM CONTENT OF MINERALS AND ROCKS (IN P.P.M.)[1]

Sample	*Chromatographic*	*Spectrographic*
Muscovite (Hte-Vienne)	50	44
Albite (Madagascar)	10	12
Muscovite (Madagascar)	50–75	70
Pegmatite (Madagascar)	50	62
Pegmatite (Madagascar)	20	18
Pegmatite (Madagascar)	10	12
Pegmatite (Madagascar)	10	11
Pegmatite (Madagascar)	5–10	10
Pegmatite (Madagascar)	5	4
Pegmatite (Madagascar)	2 (approx.)	3
Pegmatite (Madagascar)	3 (approx.)	2

[1] Reproduced with the kind consent of the author. After AGRINIER (1960).

set of standard spots ranging from 200–2 p.p.m. is obtained.

Determinations of Be in prepared samples in different laboratories yielded similar results, as is shown in Tables XIII and XIV.

In an earlier paper, AGRINIER (1957, p. 191) described a method for the determination of beryllium in minerals. In this method dissolution is effected initially with hydrofluoric and perchloric acids, later with hydrochloric acid and finally with water. The solvent used for the chromatography is acetone–HCl–water (70 : 20 : 10). The detecting reagent is quinalizarin as above.

Selenium in ores and soils (AGRINIER, 1962)

This method is aimed at producing a simpler and more rapid method with a more sensitive reagent than the methods of LEDERER and LEDERER (1957, 1st ed., p. 339) and WEATHERLEY (1956). Good results are claimed on tellurides, sulphides, sulphosalts, uranium ores and soils.

The sample is taken into solution (seemingly by sodium hydroxide fusion followed by solution in water) and chromatographed in a solvent having the composition methanol 45 ml, ethanol 45 ml, water 12 ml, hydrofluoric acid (s.g. 1.14) 3 ml.

Detection is achieved by spraying with a 3% potassium iodide solution. The R_F values and colour of spots of related elements are given in Table XV.

Selenium contents of the order of 0.1–0.2% have been detected in a number of ore minerals (see Table XVI). The selenium content is determined either by comparison with standard chromatograms or by densitometer readings.

TABLE XV

R_F VALUES AND COLOUR OF SPOTS OF SELENIUM AND RELATED ELEMENTS

Elements	R_F	*Colour*
Ge^{4+}	0.01	–
Te^{4+}	0.55	yellow brown
Se^{4+}	0.75	brown
As^{5+}	0.95	brown

TABLE XVI

SELENIUM CONTENT OF ORE MINERALS[1]

Mineral	*Locality*	*Selenium content (%)*
Bismuthinite Bi_2S_3	Tazna (Bolivia)	0.2
	Llallagua (Bolivia)	0.075
	Persberg (Sweden)	0.080
	Meymac (France)	0.050
Tetradymite $Bi_2Te_2S_3$	"Indice de Testa" (Corsica)	0.3
	Csiklova (Hungary)	0.6
Metacinnabar HgS	Guadalcazar (Mexico)	0.8
Kobellite $Pb_2(Bi,Sb)_2S_5$	Vena (Sweden)	0.08
Native tellurium	Boulder County (Colorado)	0.10

[1] Reproduced with the kind consent of the author. After AGRINIER (1962).

Molybdenum in ores, rocks and soils (AGRINIER, 1959)

The separation of molybdenum from many metals, particularly uranium, iron, titanium and chromium, is achieved by treating a solution of the sample with ammonium hydroxide and filtering. The solution of ammonium molybdate, vanadate, etc., is chromatographed in the following solvent: ammonium hydroxide (55 ml) acetone (45 ml).

Detection of molybdenum and associated metals is achieved by the use of "Tiron" (1.2-dihydroxybenzene, 3.5 disulfonate of sodium) (Tiron 2 g, water 20 ml, acetone 80 ml), but for very low concentrations a 5% aqueous solution of potassium ferrocyanide is used.

Semiquantitative estimations may be made by reference to standard chromatograms of solutions of known strength.

The paper chromatographic methods, developed at the Chemical Research Laboratory, Teddington, England (HUNT and WELLS, 1954; HUNT et al., 1955; NORTH and WELLS, 1959), for the determination of copper, cobalt, nickel, niobium, tantalum, lead, uranium and bismuth in soils represent a significant advance in geochemical prospecting. It is regrettable that the methods received extensive field use under African tropical conditions where the problem of keeping the chromatography paper in a usable condition created

TABLE XVII

R_F VALUES AND COLOUR OF SPOTS OF MOLYBDENUM AND RELATED ELEMENTS

Elements	R_F	*Coloration (if separate)*
Fe	0	brick red
U	0	brown
Cr	0	yellow
Ti	0	yellow
V	0–0.3	grey to black
Mo	0.70	yellow
Co } only in concentra-	0.90	red–brown
Cu } tions of many %	0.95	pale brown

great difficulty (C. Bursill, personal communication, 1961). In more favourable regions, the methods merit much more extensive use. A specially designed slotted paper was introduced (CRL-1 by Whatman). This permitted ten chromatograms to be "run" simultaneously. The authors found that an ascent of about 9 cm (on the paper strips between the slots) gave a sufficiently good separation of the metals in the solvent mixture used.

Separation and estimation of copper, cobalt, and nickel

The soil sample is ground and passed through an 80 mesh sieve. A weighed sample (0.5 g) is mixed with 1 g of powdered potassium bisulphate in an 18 × 150 mm hard glass test-tube. Fuse gently for 1 minute. Cool and then add 2 ml of acid mixture (prepared by mixing 50 ml HCl (s.g. 1.18), 5 ml HNO_3 (s.g. 1.42) and water to make up 100 ml). Immerse the lower inch of the tube in boiling water for 10 minutes, with occasional shaking. Allow to cool whereupon silica settles out. Take 0.01 ml of the clear supernatant liquor as the chromatographic sample and apply it to one strip of the slotted paper. Nine similar samples can be applied to the same sheet. The sheet is rolled into a cylindrical shape and held with a paper clip. Stand the cylinder in a 600 ml beaker floating in a boiling water bath (to dry the sample solution added). After three minutes transfer the paper cylinder to another 600 ml beaker, to which 10 minutes previously, 20 ml of the following solvent has been added: Ethyl methyl ketone (pure and dry) 15 ml, hydrochloric acid (s.g. 1.18) 3 ml, water 2 ml. Cover beaker with a petri dish. Develop until the solvent front just reaches the top of the strip. Remove chromatogram and allow to dry. Neutralize paper by exposure to ammonia vapour for two minutes. Spray both sides of paper with 0.1% solution of rubeanic acid (0.1 g rubeanic acid dissolved in 60 ml of warm ethanol, filtered and made up to 100 ml with water). Allow to dry and estimate metal content by comparison with standard chromatograms. Copper, cobalt and nickel form bands at positions corresponding to R_F values respectively 0.68, 0.29[1], 0.27. The copper sometimes gives a double spot. If the metal content of the sample exceeds that of the highest standard, repeat with an aliquot from the sample which is diluted with mixed HCl and HNO_3. Alternatively, a smaller soil sample (e.g., 0.1 g) is taken. Standard solutions are prepared as follows: Acid mixture: 50 ml hydrochloric acid (s.g. 1.18), 5 ml nitric acid (s.g. 1.42), make up to 100 ml with water.

No. 1 bulk solution (indicates 2.000 p.p.m. of each metal in soil): Weigh 0.01 g of each metal in the chloride form, that is, 0.0268 g of $CuCl_2.2H_2O$, 0.0405 g of $CoCl_2.6H_2O$, 0.0405 g of $NiCl_2.6H_2O$. Dissolve together in acid mixture and make up to 20 ml. A series of standard solutions can be prepared by procedures given in Table XVIII, which are suggested by the present author.

It should be made clear that these standards are so arranged in concentration that a direct reading of p.p.m. in the soil can be made from the chromatogram. They are only usable for direct comparison if 0.5 g of soil sample is made up finally to 2 ml of test solution. This treatment of the soil represents in effect a four-fold dilution. The *actual metal concentration* in the standard solutions is one quarter of that shown on the label. Hence the weighing of the 0.5 g of soil and the measurement of the 2 ml of acid mixture into which the fusion melt is finally dissolved and the measurement of the 0.1 ml

[1] From photographs in the original papers R_F for Co^{2+} appears to be about 0.5 (present author).

TABLE XVIII

PROCEDURES FOR PREPARING STANDARD SOLUTIONS

Solution p.p.m.	*Quantity of bulk solution ml*	*Bulk solution no.*	*Acid mixture*
2,000[1]	2	1	-
1,500	1.5	1	0.5
1,000	1	1	1
600	0.6	1	1.4
400	0.4	1	1.6
300	0.3	1	1.7
200	0.2	1	1.8
150	0.15	1	1.85
100[2]	1	1	19
60	2.3	2	0.8
40	0.8	2	1.2
30	0.6	2	1.4
20	0.4	2	1.6
10	0.2	2	1.8

[1] Reserve 2 ml of no. 1 bulk solution.
[2] This solution constitues no. 2 bulk solution (reserve 2 ml).

aliquot should be carried out with reasonable accuracy.

To make this quite clear, if a 0.5 g of soil sample is fused, dissolved finally in 2 ml of acid solution and a 0.01 ml aliquot is chromatographed and if the Cu spot is comparable in size and density to the 200 p.p.m. standard chromatogram (prepared by chromatographing 0.01 ml of the 200 p.p.m. standard solution), then the soil contains 200 p.p.m. of copper even though the actual metal content of the standard 200 p.p.m. solution is in fact 50 p.p.m.

To prepare "final" standard solutions that are comparable to test solutions, take 2 ml of each standard solution, prepared as described above. To each add 0.5 g potassium bisulphate and warm to dissolve; cool to allow some potassium salts to crystallise out. Stopper tightly in plastic phial. Use 0.01 ml aliquots of the supernatant liquors to prepare the standard chromatograms using the same procedure as has been given already.

Estimation of niobium

Weigh 1 g of soil (minus 80 mesh) into a 10 ml polythene beaker and add 5 ml hydrofluoric acid (40%) and evaporate to dryness on a water bath. To residue add 2 ml dilute hydrofluoric acid (10%) from polythene pipette with rubber bulb. Stir well with polythene rod and stand 30 minutes. Apply 0.05 ml of supernatant liquor to paper and allow to dry (stand 1 hour). The solvent is prepared as follows: ethyl methyl ketone (pure) 17 ml, hydrofluoric acid (40%) 3 ml. Procedure is the same as for the other metals with the exceptions: *(a)* After application of the test solutions the paper is allowed to dry for 1 hour. *(b)* The solvent is allowed to stand for 30 minutes in the covered beaker before the paper cylinder is stood in it. After neutralisation of the chromatogram with ammonia vapour, detection is achieved by spraying with tannic acid (2% aqueous solution). After drying, the test chromatogram is compared with standards. Standard chromatograms must be prepared daily (alternatively test chromatograms should be kept over night and then compared with old standards).

To prepare the standard, ignite "specpure" niobium pentoxide to 500–600° C for 30 minutes, weigh 0.400 g of the oxide into a platinum dish and warm with HF and a little nitric acid until solution is complete. Evaporate to dryness on a water bath; repeat evaporation twice with HF alone. Dissolve residue in a minimum volume of warm dilute hydrofluoric acid and cool. Dilute solution to 100 ml and transfer to a polythene bottle. Dilute portions of the solution with water containing 2 ml of 40% w/w HF to give a series of standard solutions. Aliquots of these solutions are used for the preparation of standards covering the range 0.1–8.0 μg of niobium pentoxide. These correspond to 4–320 p.p.m. of the pentoxide in soils. The range can be extended by taking smaller aliquots of the soil test solution. However, for higher-grade materials the method of HUNT and WELLS (1954) is to be preferred.

If the mineral ilmenorutile is present in the soil, a potassium bisulphate fusion is recommended, 1 g of soil (minus 80 mesh) and 4 g of potassium bisulphate being fused in a platinum crucible for 1 or 2 minutes. To the cooled melt add 5 ml of hydrofluoric acid

(10%) stir and stand for 30 minutes. Take 0.02 ml of the supernatant liquor as the test sample and proceed as described above.

Estimation of tantalum

Fusion of tantalum-bearing soils with potassium bisulphate is avoided since the later use of hydrofluoric acid produces the insoluble potassium tantalofluoride. While tannic acid is a specific reagent for niobium in the presence of tantalum, there is no specific reagent for tantalum in these circumstances. The solvent mixture is adjusted to achieve chromatographic separation of these two metals.

The procedure is the same as before except that if niobium is present, the test application should be dried for $1^1/_2$ hours if the temperature is about 25°C or 2 hours at about 20°C. The solvent mixture required is 20 ml of the mixture ethyl methyl ketone (pure) 90 ml, hydrofluoric acid (40%) 2 ml, water 8 ml. After removal, the chromatogram is sprayed immediately with quinalizarin solution (0.05 g in 10 ml pyridine then diluted to 100 ml with acetone). After spraying, expose the chromatogram to ammonia vapour for a few minutes and then expose to acetic acid vapour for a few minutes. Compare test chromatograms with standards.

Standards are prepared as follows: 0.0819 g of tantalum metal is warmed with hydrofluoric acid (40%) and nitric acid (s.g. 1.42) until completely dissolved; evaporate to dryness. The evaporation is repeated twice more with hydrofluoric acid alone. The residue is dissolved in 5 ml hydrofluoric acid (40%) and then diluted to 100 ml with water. Portions of this solution, the 1,000 p.p.m. Ta_2O_5 standard, are diluted with the requisite volumes of water containing 5 ml of hydrofluoric acid (40%) to give a series of standard solutions.

Estimation of lead

The inability of lead to form complexes in methanol-HCl as readily as other metals is used to achieve a separation.

Weigh 1 g of soil (minus 80 mesh) into a 12 mm × 100 mm rimless test tube. Add 1 ml of dilute nitric acid (1 :3) and digest on a water bath for 1 hour (only the lower one inch of the tube is immersed). Cool and allow to settle. Apply 0.01 ml aliquot of the clear supernatant liquid to the paper. Dry for 30 minutes. Chromatograph in a 600 ml beaker with 20 ml of solvent (1 ml HCl [s.g. 1.18] and 19 ml of pure methanol). After development, allow chromatogram to dry for at least 30 minutes and then expose it to ammonia fumes for 2 minutes. Stand in air for 15 minutes and then spray with a buffered solution of dithizone (few mg in 20 ml of acetone containing 1 drop 2 N NH_4OH and diluting to 100 ml with a 5% aqueous solution of ammonium acetate).

Standard solutions are prepared as follows: 0.08 g lead nitrate (containing 0.05 g lead) is dissolved in dilute nitric acid (1 :3) and made up to 20 ml. An aliquot of 0.01 ml of this solution contains 25 micrograms of lead which corresponds to 2500 p.p.m. of lead in the soil. Standards of lower concentration are prepared either by dilution (with dilute nitric acid as above) or by weighing out smaller quantities of lead nitrate.

After chromatographing 0.01 ml aliquots of standard solutions and spraying with dithizone (as above), the standard chromatograms can be stored in the dark for a few days. However, the authors recommend the use of fresh standards daily (HUNT et al., 1955).

Estimation of uranium

The procedure for uranium is hindered by the interference of phosphates present in some samples obtained from soils containing low iron and aluminium contents. To overcome this, the soil extracts are saturated with aluminium nitrate after the method of THOMPSON and LAKIN (1957).

Weigh 1 g of soil (minus 80 mesh) in a small platinum dish, add 2 ml nitric acid (s.g. 1.42) and 2 ml hydrofluoric acid (40%). Evaporate to dryness on a water bath; add 1 ml dilute nitric acid (1 :3) to the residue, stir well and leave to cool and settle. Apply 0.05 ml aliquots to the chromatography paper. Allow to dry for 1 hour. Chromatograph in a covered 600 ml beaker with 20 ml of solvent (2 ml nitric acid [s.g. 1.42], 1 ml water, 17 ml ethyl acetate A.R. grade). After development, allow the solvent on the paper to evaporate and then spray the sheet with 5% aqueous solution of potassium ferrocyanide. Dry and compare with standard chromatograms.

Standard solutions are prepared by dissolving uranyl nitrate, $UO_2(NO_3)_2.6H_2O$, in dilute nitric acid (1:3). A solution containing 2,000 p.p.m. is prepared by dissolving 0.0844 g of uranyl nitrate in dilute nitric acid (1 :3) and making up to 20 ml. Other standard solutions are prepared by dilution of the 2,000 p.p.m. standard or by weighing and dissolving smaller fractions of the uranium nitrate. Aliquots of the standard solutions to be chromatographed should be in this case the same as that of the soil test solution, that is, 0.05 ml.

Estimation of bismuth

The organic matter present in the soils interferes with the movement of bismuth on the chromatogram. Hence the soils were preheated to dull red heat for a few minutes to burn off the organic matter. The procedure used (with modifications by the present author) is as follows: weigh 1 g of soil (minus 80 mesh) into a crucible and heat to a dull red heat for a few minutes. Cool and transfer residue to a test tube. Add 1 ml of dilute nitric acid (1 :3) and digest on a hot water bath for 15 minutes with only the lower inch of the tube immersed. Remove the tube, cool, make up volume to 1 ml with dilute nitric acid and allow insoluble matter to settle. Apply an aliquot of 0.01 ml of the clear supernatant liquid to the chromatography paper and dry at least 15 minutes. Then stand for a further 15 minutes over a saturated aqueous solution of ammonium sulphate. The development is effected in 20 ml of solvent (0.5 ml nitric acid [s.g. 1.42], 0.5 ml water, 19 ml of dry dioxan [A.R.]) in a covered 600 ml beaker. After development, the solvent on the chromatogram is allowed to evaporate. Then the chromatogram is sprayed with a solution of 0.2% dimercaptothiodiazol in *N* sodium hydroxide. Estimation of bismuth is made by comparison with standard chromatograms.

Standard solutions are prepared by dissolving a bismuth salt in dilute nitric acid (1 :3). If 0.064 g of bismuth chloride ($BiCl_3.H_2O$) is dissolved in dilute nitric acid and the volume made up to 20 ml, a standard solution of 2,000 p.p.m. of bismuth is obtained. Other standard solutions are made by dilution of this solution or by special preparation. The standard chromatograms fade slowly when exposed to sunlight. Fresh standards are needed every two or three days.

Determination of lithium in minerals

AGRINIER (1957a, p.184) has described also a method for the determination of lithium in minerals. Magnesium and manganese are shown to interfere with lithium (having about the same R_F in the solvent used) but the method proposed is claimed to eliminate the effect of interfering metals. The procedure is: Grind the sample finely and fuse 5–10 mg in a platinum crucible with 50 mg sodium carbonate for 5 minutes. Dissolve the residue in 2 ml water and filter through a fast paper in a micro-funnel (1.5 cm diameter). Oxides of iron and magnesium are removed by this step. Collect the filtrate in a 2 ml micro-beaker and evaporate slowly to dryness. Add 10 *N* HNO_3 (1 ml) and evaporate to dryness; repeat with 12 *N* HCl (1 ml). Dissolve residue in 1 ml water.

Apply 0.01 ml of sample solution to the paper and chromatograph in ethanol–methanol–ammonium hydroxide (49 :49 :2). An ascent of 16 cm is recommended. After development, dry the chromatogram in a current of cold air. Spray with a reagent prepared as follows: dissolve 3 g silver nitrate and 200 mg alizarin in 5 ml water; make up to 100 ml with ethanol; filter through a rapid paper. After spraying, dry the chromatogram and heat under an infra-red lamp. The R_F values of lithium and associated ions are:

Li^+	0.625
NH_4^+	0.40
Na^+	0.26
K^+	0.11
Mn^{2+}	0.60
Mg^{2+}	0.59

Lithium was detected in lepidolite, amblygonite, lithiophylite and spodumene.

Determination of boron in minerals

The difficult task of proving the presence of boron in minerals (especially in tourmalines) is overcome by the following paper chromatographic method proposed by AGRINIER (1957a, p. 188).

Grind finely 5–10 mg of the sample. Prepare a sodium carbonate bead (about 25 mg) on platinum wire. Take up the sample on the bead and heat for 5 minutes in a bunsen flame (oxidising). Cool and then crush the bead. Place in a 1 ml microbeaker, add 0.5 ml water and warm for 1 minute.

Apply 0.01 ml of sample solution on the paper (no. 1 Whatman) and, after drying, chromatograph in ammonium hydroxide (100 ml) in 10 cm × 10 cm covered vessel. After an ascent of 10 cm is achieved, take out and dry the chromatogram. Spray with a reagent prepared as follows: dissolve 3 g silver nitrate and 200 mg alizarin in 5 ml water, make up to 100 ml with ethanol and filter. The reagent is stable for about 10 days but it is usable for a longer period if refiltered from time to time and stored in a dark glass bottle. After spraying, dry the chromatogram for 1 or 2 minutes then expose it to infra-red light. After some minutes of this treatment, boron gives a greyish spot that turns brown if heated. The R_F values of boron and associated ions are given in Table XIX.

TABLE XIX

R_F VALUES OF BORON AND ASSOCIATED IONS

Elements	R_F	*Colour*
Na^+	0.40	brown spot
BO_4^-	0.70	pale violet spot that turns grey
Na_2CO_3	1.0	brown spot

Agrinier reports that circular chromatography gives satisfactory results, the boron in tourmaline being characterised by a medial aureole.

Determination and semi-quantitative estimation of silver in silver minerals and argentiferous galenas (AGRINIER, 1957b, p. 277)

Agrinier recommends the following procedure: for silver minerals (1% Ag or more) dissolution is achieved as follows: weigh

10 mg of the finely ground mineral and heat gently in a silica micro-crucible until completely fused. Cool, take up in 10 *N* HNO_3 (1 ml) and dry. Repeat this operation twice more. Dissolve the residue in 10% HNO_3 (1 ml). Stir and allow to settle. For argentiferous galenas (Ag 300 p.p.m.) a larger sample is needed. Grind finely and melt 1 g galena, dissolve twice in 10 *N* HNO_3 (2 ml) bringing to dryness each time. Dissolve in 10% HNO_3 (1 ml).

Apply 0.01 ml of the sample solution to the paper and develop the chromatogram using 100 ml ammonium hydroxide in a sealed vessel. An ascent of 10 cm is recommended. Upon removal from the vessel, dry the chromatogram in a current of cold air. Spray with reagent (saturated solution of p-dimethylaminobenzylidene rhodanine in acetone). To remove excess reagent, wash the chromatogram twice in acetone. Dry and compare with standard chromatograms.

Standard solutions are prepared as follows: weigh 1 g of silver metal and dissolve it in 10 *N* HNO_3 (20 ml). Evaporate to dryness on a sand-bath. Dissolve the residue in 10% nitric acid (100 ml). This constitutes a 1% silver solution. Prepare the following dilutions: 0.8, 0.6, 0.5, 0.4, 0.3, 0.2, 0.10, 0.05 and 0.03%.

Standard chromatograms are prepared by applying 0.01 ml of each of the diluted solutions at suitable spaces on the paper (large size Whatman CRL-1 type). Develop immediately in ammonium hydroxide solvent which has been allowed to stand in the closed vessel for 30 minutes before the introduction of the paper. An ascent of about 10–15 cm suffices. Remove the chromatogram and allow to dry. Spray with the reagent, dry and wash twice with acetone.

The percentage content of silver in the silver mineral is obtained by multiplying the value obtained by comparison by one hundred.

In the case of the argentiferous galenas, since the sample taken is 1 g, the percentage content of silver is obtained directly from comparison with the standard chromatograms.

Determination and semi-quantitative estimation of nickel, cobalt and copper in ores (AGRINIER, 1957b, p. 279)

For these separations and estimations, Agrinier recommends

vessels 25 cm × 9 cm. Whatman no. 1 paper is preferred to that of d'Arches (a French paper). Careful sorting of the crushed ore under a low power stereomicroscope is recommended to avoid the incorporation of extraneous material in the sample. Thereafter the following procedure is followed: grind the sorted ore finely and place 10 mg of it in a silica microbeaker. Add 10 *N* HNO_3 (3 drops) and evaporate to dryness. Dissolve the residue in *N* HCl (3 drops). Repeat this operation twice, drying gently each time. Dissolve the residue in 15% HCl (2 drops). The solvent mixture is prepared as follows: acetone 87 ml, 12 *N* HCl 8 ml, water 5 ml. The chromatogram is developed until an ascent of 10 cm is achieved and is then dried in a current of cold air. The chromatogram is then exposed to ammonia fumes for five minutes. Spray both sides of the chromatogram with rubeanic acid (1% solution in ethanol). Wait several minutes for spots to appear. If it is desired to determine zinc as well, 1% diphenylcarbazide solution is added to the reagent. Identification is based on the data, given in Table XX.

TABLE XX

R_F VALUES OF NICKEL AND RELATED ELEMENTS

Element	R_F	*Colour*
Ni^{2+}	0.05	blue spot
V^{3+}	0.2	violet spot
Co^{2+}	0.5	brownish yellow spot
Cu^{2+}	0.65	olive green spot
Zn^{2+}	0.90	rose spot

Estimations are made by comparison with prepared standard chromatograms. Prepare standard solutions as follows: weigh 1 g Co_2O_3, 1 g CuO, 0.91 g NiO and place in a 50 ml beaker and add 12 *N* HCl (20 ml). Dry carefully without spitting. Take up the residue in 15% HCl (71 ml). This solution contains 1% of each metal per ml. Make the following dilutions: 0.8, 0.6, 0.5, 0.4, 0.3, 0.2, 0.15, 0.10, 0.05 and 0.01%. Apply 0.01 ml of each solution at suitable positions on the paper. All developments are achieved under humid conditions by placing 100 ml of water in the vessel. Thirty minutes before the separation the solvent is added to a dish which rests on the floor of the vessel. The paper as a cylinder is allowed to stand in the solvent in the dish. An ascent of 10 cm is sufficient. After development, the chromatogram is treated with reagent as above. The percentage content of the particular metal in the ore is obtained by multiplying the value obtained by comparison by one hundred.

Determination and semi-quantitative estimation of niobium, tantalum and titanium in ores (AGRINIER, 1957b, p.284)

Agrinier modifies the method of HUNT et al. (1955) by using HF–$HClO_4$ to dissolve the minerals and by chromatographing in different solvents. The following procedure is recommended for tantalum: grind the mineral finely (minus 150 mesh) and weigh 10 mg into a platinum crucible. Add 40% hydrofluoric acid (2.5 ml) and perchloric acid (2.5 ml) and evaporate gently to dryness without calcining. Dissolve the residue in 10% hydrofluoric acid (1 ml) stirring with a polythene rod for several minutes. Allow to stand for 30 minutes.

The chromatography is effected in a desiccator containing acetone (100 ml) to saturate the atmosphere therein. The solvent is placed in a polythene dish. For tantalum, the solvent is prepared 30 minutes in advance in polythene vessels as follows: acetone 90 ml, water 8.5 ml, 40% HF 1.5 ml. Stir well with a polythene rod. Apply 0.01 ml of the sample solution(s) on the paper (Whatman no. 1, large CRL-1 type) and dry well with a current of cold air for ten minutes. Clip the paper in the form of a cylinder, stand it in the solvent and close the desiccator. A 10 cm ascent is sufficient development. Prepare a detecting reagent by dissolving 10 mg of quinalizarin in 10 ml of pyridine, making up the volume to 100 ml with acetone. Filter. Spray the chromatogram on both sides as soon as it is removed from the vessel. Stand the chromatogram for 3 minutes in a desiccator containing ammonium hydroxide. Transfer for 5 minutes to a desiccator containing acetic acid. The last two operations are to intensify the coloration and to remove the excess reagent. A violet-pink spot at the solvent front is due to tantalum. At about R_F 0.1, a violet spot for niobium occurs. Below the niobium spot another violet spot indicates titanium. Iron gives a blue spot at the point of application. The coloration due to metals other than tantalum fades quickly. The latter is stable. Estimation is made by comparison with standard chromatograms prepared from solutions of known concentration.

For niobium and titanium, the solvent (prepared 30 minutes in advance) is acetone 90 ml, 12 N HCl 1 ml, 40% HF 2 ml and water 13 ml. The same chromatographic technique as described for tantalum is used. After development and neutralization with ammonia fumes for five minutes, the chromatogram is sprayed on both sides with tannic acid solution prepared as follows: dissolve 2 g tannic acid in 5 ml water and make up to 100 ml with acetone. After spraying dry and compare with the standard chromatograms.

Niobium forms an orange yellow band at the solvent front (R_F about 1.0). Titanium forms a yellow band at R_F approximately 0.70. A violet spot at R_F about 0.40 indicates iron.Other associated metals have R_F values below 0.40 and even if they give colorations these fade quickly.

Standard solutions are prepared as follows: weigh 250 mg of powdered niobium and 520 mg of titanium dioxide. Place in a platinum crucible with 40% hydrofluoric acid (6 ml) and 60% perchloric acid (4 ml) and evaporate slowly to dryness. Take up the residue in 4% hydrofluoric acid (6 ml) and evaporate to dryness. Repeat this operation twice. Dissolve the residue by warming gently with 40% hydrofluoric acid (3.75 ml). Pour the solution into a polythene flask containing 11.25 ml water. Wash the platinum crucible twice with 5 ml quantities of water which are added to the flask to make up 25 ml of solution. This is a 1% solution of each metal. Dilution of aliquots of this solution produce 0.8, 0.6, 0.5, 0.4, 0.3, 0.2, 0.15, 0.10, 0.05 and 0.01% solutions. The standard tantalum solution is prepared as follows: weigh 250 mg of powdered tantalum and treat initially with 40% hydrofluoric acid (3 ml) and perchloric acid (2 ml) and then follow the procedure given for niobium and titanium. Add 0.01 ml of each standard solution to the paper and dry at 30° C for 5 minutes. Chromatograph as for the sample solutions. Remove chromatogram after a 10 cm ascent, dry for a few minutes and neutralize in ammonia vapour. Spray both sides with tannic acid reagent and dry. For the tantalum chromatographs, spray upon removal from the vessel and then neutralise in ammonia fumes for three minutes.

The percentages of each metal in the sample are obtained by multiplying by 100 the values on the comparable standard chromatograms.

Semi-quantitative determination of arsenic in minerals

A brief outline of a method to determine arsenic semi-quantitatively has been given by AGRINIER (1961a). Minerals with arsenic in the As^{5+} state - arsenates, such as mimetite, $Pb_5(AsO_4)_3Cl$ - are dissolved in N HNO_3. Minerals with arsenic in the As^{3+} state - arsenites, such as dufrenoysite, $2\,PbS.\,As_2S_3$ - are dissolved by warming in a 10% sodium hydroxide solution and by reducing the bulk of the solution somewhat during heating until solution is complete. Then 10 N HNO_3 is added slowly until the alkali is neutralised.

The solvent used is acetone 50 ml, methyl ethyl ketone 40 ml, HNO_3 (s.g. 1.33) 10 ml, water 10 ml. Detection is achieved by spraying with a 30% aqueous solution of potassium iodide. Data on R_F values and colour of spots for arsenic and associated metal ions are given in Table XXI. The As^{3+} ion is detected with the reagent sodium hypophosphite (aqueous solution).

Semi-quantitative estimations are made for each arsenic ion by comparison with standard chromatograms prepared from solutions of known concentration. In table XXII, some results obtained by chromatography are given together with those obtained spectrographically. It has been suggested by Agrinier that the method is applicable to the estimation of mercury especially if diphenylcarbazone is used as the detecting reagent.

Semi-quantitative determination of bismuth in minerals

The method outlined by AGRINIER (1961b) is based on the

TABLE XXI

R_F VALUES AND COLOUR OF SPOTS OF ARSENIC AND RELATED ELEMENTS

Ion	R_F	*Colour*
Fe^{3+}	0.05	brown
Cu^{2+}	0.08	brown
Pb^{2+}	0.08	yellow
Ni^{2+} – Co^{2+}	0.08	–
As^{3+}	0.25	–
Ge^{4+}	0.30	–
SeO_3^{2-} – TeO_3^{2-}	0.35	brown
P^{5+}	0.54	–
As^{5+}	0.57	brown
Bi^{3+}	0.77	orange
UO_2^{2+} – Th^{4+}	1.0	–
Hg^{2+}	1.0	–

movement of bismuth nitrate on a paper chromatogram in a solvent composed of acetone 88 ml, water 12 ml, 40% hydrofluoric acid 3 ml, 10 *N* HCl 0.5 ml. The finely ground mineral, it is assumed, is taken up in nitric acid. After development, the ions are detected by spraying with a 0.2% solution of Bismuthiol (dimercaptothiodiazol, 0.1% solution in ethanol). A second spraying with ammonium polysulphide is made immediately.

TABLE XXII

ARSENIC CONTENT OF SOME ORE MINERALS[1]

Minerals	*Locality*	*Percentage arsenic*	
		Chromato-graphic	*Spectro-graphic*
Chalcolite $Cu(UO_2)_2(PO_4)_2$ + water	Limouzat	0.20	0.25
Tetrahedrite $(Cu,Fe)_{12}.Sb_4S_{13}$	Kapnik	0.75	
Polybasite $(Ag,Cu)_{16}Sb_2S_{11}$	Cornouailles	0.15	0.15
Descloizite $(Zn,Cu)Pb(VO_4)OH$	Southeast Africa	0.75	0.80

[1] After AGRINIER (1961a). Reproduced with the kind consent of the author.

TABLE XXIII

R_F VALUES FOR BISMUTH AND ASSOCIATED METALS

Metals	R_F	*Colour with Bismuthiol*	*Colour with ammonium sulphide*
Ni–Co	0.1	–	brown
Pb	0.2	–	brown
Cu	0.25	bright yellow	brown
Fe	0.60	–	brownish–green
Sb^{3+}	0.80	–	orange
Bi	0.90	orange–yellow	brown–black
Mo	0.95	–	brown–red
Hg^{2+}	1.0	–	black
Sb^{5+}	1.0	–	orange

The R_F values for bismuth and for metals commonly associated with it are given in Table XXIII.

Estimation of bismuth is made by comparison with standard chromatograms. The orange–yellow colour given by bismuth with bismuthiol is specific for that metal. The ammonium polysulphide augments the colour intensity and extends the range of detection from 10 μg down to 5 μg.

TABLE XXIV

BISMUTH CONTENT OF ORE MINERALS[1]

Mineral	*Locality*	*Bismuth content*	
		Chromatographic	*Spectrographic*
Arsenopyrite FeAsS	Puy de Dome	0.3%	0.3%
Arsenopyrite FeAsS	Finisterre	$150\text{–}200 \cdot 10^{-6}$	$200 \cdot 10^{-6}$
Tennantite $(Cu,Fe)_{12}As_4S_{13}$	Peru	0.6%	0.5–1%
Smaltite $(Co,Ni)As_3$	Saxony	$300 \cdot 10^{-6}$	$320 \cdot 10^{-6}$
Chalcolite $Cu(UO_2)_2(PO_4)_2$ + water	Haute-Loire	$350 \cdot 10^{-6}$	$330 \cdot 10^{-6}$
Schwatzite $(Cu,Fe)_{12}Sb_4S_{13}$	Tyrol	$500 \cdot 10^{-6}$	$480 \cdot 10^{-6}$

[1] After AGRINIER (1961b). Reproduced with the kind consent of the author.

To test the method, some chromatographic estimations have been compared with values obtained on identical specimens by spectrographic methods. The results are set out in Table XXIV.

In the study of the minerals cited, care was taken to show, by mineragraphic examination, that native bismuth and bismuthinite were not present as small inclusions.

Separation and estimation of selenium in silicate materials

A paper chromatographic method reported by WEATHERLEY (1956) for the separation and estimation of selenium from numerous metals in artificial mixtures "shows potentialities for geochemical prospecting".

Fuse 0.5 g of the artificial silicate material (0.35% selenium in glasses containing cobalt–nickel, chromium–iron and molybdenum–cadmium constituents) with 1 g sodium carbonate (anhydrous) in small covered platinum dish. Add dilute hydrochloric acid (5 : 3 v/v) (5 ml) to disintegrate, melt and digest on a water bath for 30 minutes. Allow to stand. Apply 0.05 ml aliquots to the paper (Whatman CRL-1 type) and allow to dry. Chromatograph in a solvent consisting of ethyl methyl ketone 80 ml, 40% hydrofluoric acid 20 ml. After development, dry the paper and spray it with thiourea (3% w/v solution in *N* HCl. Dry again and compare with standard chromatograms. Selenium advances with the solvent front (R_F 1.0).

Standard chromatograms are prepared as follows: dissolve 0.2190 g of pure anhydrous sodium selenite in 10 ml of water, add 10 *N* HCl (10 ml) and make up to 100 ml accurately. This solution contains 0.0025 mg (or 2.5 μg) of selenium per 0.05 ml aliquot. Apply suitable aliquots to give amounts of selenium from 5–45 μg to the paper. Chromatograph and spray as for the artificial silicate sample. Standard chromatograms, stored between glass plates in the dark, will keep for three months.

It is claimed that the "accuracy and reproducibility are considered as good as those by conventional gravimetric procedures, but the time involved is considerably less...". The results quoted in the original paper justify the claim.

PROSPECTING FOR FINE GOLD

The loss of fine gold during panning and other gravity methods is a problem in localities where this kind of gold occurs. The losses occasioned by the crude methods are accentuated by losses due to adsorption of gold onto colloidal particles if muddy water is used. Thus the development of a chromatographic method (NEVILL and LEVER, 1959) is particularly welcome. The method is applicable to oxidized samples (soils, gossan, eluvial and alluvial deposits) and to sulphide ores that have been suitably treated. In the latter case, about 10 g of finely ground sample is lightly consolidated in a measuring cylinder to ascertain the volume. The sample is placed in an open dish and roasted with constant rabbling for one hour at 550° C. Excess iron in the roasted ore is removed by washing and decanting three times with hot concentrated hydrochloric acid and then three times with water. The residue is dried and returned to the measuring cylinder where the original volume is restored by adding and mixing barren, finely ground material.

The usual paper chromatographic equipment is employed. In addition, a small aluminium frame is used to measure samples by bulk. The outside dimensions of the frame are $1^3/_4'' \times 2'' \times {}^1/_{32}''$ (thickness). An opening $1^1/_4'' \times 1^1/_2''$ is made in a symmetrical position. Plates of glass $2'' \times 2'' \times {}^1/_4''$ and some 2″ spring back clips are required also. Then the following procedure is followed: grind the sample to minus 65 mesh (or finer) avoiding contamination with metallic iron (remove by use of strong magnet, if necessary). Use Whatman no. 3MM 1.5″ roll chromatography paper cut into pieces 7″ long. Mark pencil lines at 1″, $3^1/_4''$ and $4^1/_2''$ from one end. Place the aluminium frame on the paper at the 1″ mark. Fill the frame with the sample, which is compressed to remove excess pores and smoothed off level with a spatula. This leaves about 1.4–1.5 g of sample on the paper after the frame has been removed. Clamp $2'' \times 2''$ plate glass squares on both sides with spring clips. The paper with its attachment is suspended vertically with the lower (one inch) end dipping into dilute aqua regia ($HCl : HNO_3$: water = 3 : 1 : 4) to a depth of ${}^1/_4''–{}^3/_8''$. Allow the acid to rise through the paper and sample to the $3^1/_2''$ mark (usual duration about 30 minutes). Remove from acid and take off clips and glass. Dry paper strip in an oven at 55–65° C for 15 minutes and then brush as clean as possible. Heat again in oven for 10 minutes. Brush completely clean. The gold, at this stage, is contained in the paper.

Develop the paper by the ascending technique in a solvent consisting of ethyl acetate 85 ml, nitric acid (s.g. 1.42) 10 ml, water 5 ml, potassium chlorate 0.2 g (dissolve in acid first). Ascent to the $4^1/_2''$ mark takes about 20 minutes. After development, the chromatograms are dried for a few minutes and then sprayed with freshly made stannous chloride solution (in 5 N HCl). The gold advances with the solvent front and is detected as a brown or purple line. Warm gently to intensify the coloration due to the gold but not to produce the strong colours due to iron. If the gold develops a comet, repeat using a half frame or even quarter frame of sample. Estimation of the gold content of the sample is made by comparison with standard chromatograms. Artificial standards from one to ten pennyweights per ton (approximately 1.5–15 p.p.m.), are prepared and chromatographed.

The method was tested in field trials which were carried out on samples obtained from battery tailings, shaft dumps and from

TABLE XXV

ESTIMATIONS OF GOLD CONTENT IN OXIDIZED AND SULPHIDE ORE[1]

Sample no.	*Chromatographic estimate dwt./ton*	*Fire assay estimate dwt./ton*
1 oxidized	3.0	2.99
2 oxidized	5.0	5.20
5 oxidized	20.0	22.00
average of ten	1.4	0.45
16 sulphide	16.0	16.50
17 sulphide	1.0	1.40
18 sulphide	4.0	3.35

[1] After NEVILL and LEVER (1959). Reproduced with permission of the authors

trenches. Double samples were taken, one to be analysed by fire assay and the other chromatographically. Some results are shown in Table XXV.

Sulphide samples were not roasted in the field but it should not be difficult to carry out this step under such conditions.

ANALYSIS OF COAL ASH

Some troublesome operations, met with in the classical methods of coal ash analysis (KRAUS and MOORE, 1950), are avoided if an ion-exchange method (ELLINGTON and STANLEY, 1955) is used.

A wet oxidation method (with a nitric acid-sulphuric acid mixture) described by ELLINGTON and ADAMS (1951), is used to take the inorganic constituents of coal into solution. Silica remains as an insoluble residue and most metals, except the volatile ones, go into solution. As a final preparatory step, the solution is diluted with water so that the sulphuric acid strength is not more than 0.4 *N*.

A first column (20 cm × 1 cm) is packed with amberlite IR-120 (H) and a second (30 cm × 1 cm) with amberlite IRA-400 (Cl). A third column (40 cm × 1 cm) is packed also with the IR-120 (H). The present author has used Zeo-Karb 225 (H) and De-Acidite FF (Cl) respectively as substitutes for the two resins mentioned.

The main procedure is set out in Table XXVI, which is only a slight modification of the original.

After the metals have been separated by displacement elution from the resins, some choice of methods of estimation of each metal is available. The present author has used the following methods.

Iron was determined gravimetrically as ferric oxide after precipitation as the hydroxide from the chloride solution. Sodium, potassium and calcium were determined by normal flame photometric methods. Aluminium was dried together with citric acid, which was burnt off in a muffle furnace, the aluminium being determined gravimetrically as the oxide.

Magnesium was determined on a starch column by the method of DYKYJ and CERNY (1945), (see pp. 43–44):

TABLE XXVI

SEPARATION OF INORGANIC CONSTITUENTS OF COAL WITH ION EXCHANGE RESINS[1]

Solution containing cations in 100 ml 0.4 N H_2SO_4	*Column A*
Pass onto column A. ⟶	Amberlite IR-120 (H). *(a)* Wash with 100 ml water. Effluent contains H_2SO_4 only and is rejected. *(b)* Wash with 100 ml 3 *N* HCl. Evaporate effluent down to 3 ml, add 5 ml 10 *N* HCl. Pass onto column B.
Column B ⟵	
Amberlite IRA-400 (Cl). Wash with 40 ml 8 *N* HCl.	
Resin. Wash with 0.1 *N* HCl (100 ml). Effluent contains Fe.	Effluent. Evaporate to dryness. Dissolve in 20 ml water. Pass onto column C.
Column C ⟵	
Amberlite IR-120 (H).	
(a) Wash with 0.4 *N* HCl.	Effluent between 200 and 340 ml contains Na. Effluent between 540 and 820 ml contains K.
(b) Wash with 5% citric acid buffer for which pH is adjusted to 3 by adding ammonia.	Effluent between 0 and 240 ml contains Ti. Effluent 240–600 ml reject. Effluent between 600 and 900 ml contains Al.
(c) Wash with *N* HCl.	Effluent between 40 and 100 ml contains Mg. Effluent between 360 and 680 ml contains Cu.

[1] After ELLINGTON and STANLEY (1955).

A 10 cm column of raw starch is developed, in a 50 ml burette above a glass wool pad, by sedimentation from an aqueous suspension. A very small drop of methylene blue is added to the supernatant liquid which is then drained slowly from the burette. The methylene blue is adsorbed on the uppermost starch. The magnesium solution (as the chloride) is passed through the column, the position of the blue ring being measured carefully before and after. The passage of the solution is slow, taking about 30 hours, but requiring no attention during that time. The column is washed with water and another blue ring is introduced. A magnesium chloride solution (60 ml) of known strength is then passed through the column, the displacement of the ring being noted. The displacement is proportional to the mass of magnesium passing and is independent of the concentration. Thus

$$M_U = \frac{M_K \cdot D_1}{D_2}$$

where M_U = mass of magnesium in sample, M_K = mass of magnesium in known solution, D_1 = distance moved by 1st ring, D_2 = distance moved by 2nd ring.

Titanium was determined colorimetrically as follows. Reduce the 240 ml citric acid solution (containing the titanium) to less than 50 ml on a water bath and then make up to 50 ml with water. Add 20% hydrogen peroxide solution (20 ml accurately). Increase the volume to 250 ml by the addition of 5% sulphuric acid. The determination is made on a colorimeter by reference to standard graphs obtained from known solutions.

IDENTIFICATION OF METAL IONS IN MINERALS AND MINERAL IDENTIFICATION

Formerly, the method known generally as "blowpipe analysis" was used with considerable success in both the laboratory and field to identify metals in minerals. However, in the laboratory, it was superseded by the introduction of improved microscope equipment and techniques and by the availability of equipment for X-ray, spectro-chemical and thermal analyses. The declining importance of the blowpipe method in the laboratories has led to its virtual disappearance from the curricula of universities. Thus the skill of former exponents of the method has not been inherited by the present field geologists. They, in turn, scorn the method on the false grounds of its unreliability. The paper chromatographic methods of identification of metal ions in minerals have many advantages over the blowpipe methods which now must be regarded as obsolete.

In contrast to the large number of chromatographic methods

designed specifically to determine one metal (and sometimes a few metals) in minerals, the author has described a general qualitative scheme for the identification by paper chromatography of more than fifty metal ions (RITCHIE, 1961). Until recently, a general scheme for the identification of metal ions in minerals has been restricted by the problem of dissolution of the rocks or minerals. Hence, the publication of a scheme for the identification of metal ions in minerals was restricted to the opaque ore minerals (RITCHIE, 1962) since these minerals offer no great problems of dissolution.

The highly succesful use of sodium peroxide fusion in zirconium crucibles (BELCHER, 1963) now provides a unique universal method for the dissolution of minerals. It permits the application of the general paper chromatographic scheme of the author to the problem of identification of metal ions in all minerals.

By the kind permission of the editors of the *Journal of Chemical Education* and *Economic Geology* it has been possible to regroup the essential parts of the author's publications in those journals into a more comprehensive scheme.

Dissolution of the minerals

General methods of preparation of test samples of geological material have been discussed (see pp. 33–39). The sodium peroxide fusion method is the best for the dissolution of minerals generally. Of course, some minerals are readily soluble in acids (see Tables III and IV) and can be taken into solution by them. For semi-quantitative methods, a weighed sample is taken and made up to a convenient final volume.

Solvent mixtures

A small number of metal ions which are difficult to separate from each other makes it necessary to use four solvent mixtures to achieve general separation. However, when only a few metals are contained in the sample, separation can be achieved in only one or two solvent mixtures. Since two of the mixtures contain hydrofluoric acid, vessels and apparatus should be of polythene. For the sake of uniformity, and because of its durability under field conditions, polythene is recommended wherever possible. The solvent mixtures are placed in wide-mouthed jars of about 6 litres capacity and having screw-top or clamp-on polythene lids.

Solvent 1

Development time: 6 hours.

Effective life if well sealed: 4 or 5 days.

Butanol 50 ml, 10 *N* HCl 25 ml, 40% HF 1 ml and water 24 ml.

Solvent 2

Development time: 6 hours.

Effective life if well sealed: 4 or 5 days.

Add to separating funnel — butanol 50 ml, water 50 ml and 40% hydrobromic acid 5 ml. Shake well for one or two minutes and allow the phases to separate. Run off and discard the lower (aqueous) layer. To the butanol fraction add 40% HBr (20 ml).

Solvent 3

Development time: 3 hours.

Effective life if well scaled: 3 or 4 days.

Ethanol 30 ml, methanol 30 ml and 2 *N* HCl 40 ml.

Solvent 4

Development time: 1 hour.

Effective life if well sealed: 1 or 2 days.

Acetone 90 ml, 10 *N* HCl 5 ml, 40% HF 1 ml and water 4 ml.

If large numbers of determinations are to be made, double quantities are more suitable. Since hydrobromic acid reacts to some extent with polythene (ALLEN and PICKERING, 1961), the jars should be permanently numbered and used on all occasions for the same solvent.

Preparation of chromatograms

The data to be quoted for this method were obtained on Whatman no. 1 chromatography paper at 21°C±1°C. Similar data (but not precisely the same) could be obtained on other papers and at other temperatures.

Thin strips of paper (about 3 cm × 25 cm) may be used but larger rectangles (about 30 cm × 25 cm), upon which parallel chromatograms may be "run", are usually more effective.

Since the method is qualitative, the test solution is applied to the paper from a thin glass capillary tube in the form of a half-inch smear. The applications are allowed to dry. The narrow strips are suspended from hooks in the lid or the larger rectangles are converted into slightly gaping cylinders with a paper clip at the top. The jar is sealed and covered by a second larger vessel to minimise unequal heat or air currents. In using the general scheme in the search for quite unknown metal content, the numerous detection tests require that four chromatograms of each test solution be prepared in each solvent. Parallel chromatograms, using known solutions of metals alongside the unknown often eliminate further search for the identity of the metals in the sample solution.

The development times stated normally allow approximately a 15 cm ascent. Inspection during development is possible but should be avoided.

Detection of ions on chromatograms

For a complete analysis, the four chromatograms from each solvent are tested as shown below. It will be clear that more than one reagent can be used on a chromatogram.

Chromatogram 1. $(NH_4)_2S_x$ test; $SnCl_2$ test; $SnCl_2$ + KCNS + HCl test.

Chromatogram 2. 8-hydroxyquinoline (pH 2–6) test.

Chromatogram 3. 8-hydroxyquinoline (pH 6–10) test; sodium rhodizonate test.

Chromatogram 4. Chlorine water test; p-dipicrylamine test.

Data relating to coloured complexes in the four solvents are given in Table XXVII. Readers are referred to the precautions to be adopted in interpreting these coloured spots on the chromatograms (see pp. 54–55).

Coloured sulphides are detected on one chromatogram from each solvent by holding it over a concentrated solution of ammonium polysulphide or by dipping it into a dilute aqueous solution (LEDERER and LEDERER, 1957, p.481). Numerous ions, which can be detected positively in this way, are shown in Table XXVIII.

The formation of *coloured or fluorescent insoluble hydroxyquino-*

TABLE XXVII

COLOUR OF SPOTS ON CHROMATOGRAMS AS TAKEN FROM SOLVENTS[1]

Ion	*Solvent 1*	*Solvent 2*	*Solvent 3*	*Solvent 4*
Bi^{3+}	–	–	yellow	brick-red
Cd^{2+}	–	–	yellow	yellow
Au^{3+}	yellow	yellow	yellow	yellow
Pt^{4+}	–	–	–	brown
Pd^{2+}	–	brown	brown	brown
Fe^{3+}	yellow	yellow	yellow	yellow
Cu^{2+}	blue–green	purple	blue–green	green
Co^{2+}	pink	pink	pink	pink or blue
Ni^{2+}	green	yellow	yellow–green	green
Mn^{2+}	brown	brown	brown	–
Cr^{3+}	green	green	green	green
V^{5+}	green	green	yellow–brown	green
MoO_4^{2-}	–	blue	–	blue
Ti^{4+}	–	–	yellow–brown	–

[1] After RITCHIE, 1961. For preparation see p.102.

TABLE XXVIII

DETECTION BY $(NH_4)_2S_x$ SOLUTION

Ion	*Colour of spot*	*Ion*	*Colour of spot*
Ag^{+}	black	As^{3+}	black or yellow
Hg^{2+}	black	MoO_4^{2-}	yellow
Pb^{2+}	orange to brown–black	Sb^{3+}	yellow[1]
Tl^{+}	dark brown	Sn^{2+}	brown–black
Tl^{3+}	brown		
Mn^{2+}	brown–black	Fe^{3+}	black
Cu^{2+}	dark brown	Co^{2+}	brownish–black
Cd^{2+}	yellow	Ni^{2+}	brownish–black
Bi^{3+}	brown		

[1] Usually soluble in alkaline solutions. Easily detected otherwise.

lates of metals is well known (WELCHER, 1947; STEVENS, 1959). For our purpose, two hydroxyquinoline solutions of different pH ranges are used. Spraying the dipped chromatograms with ammonia produces further pH variation which sometimes augments or sometimes diminishes the fluorescence.

The first solution is prepared as follows: dissolve about 2 g of 8-hydroxyquinoline in some ethanol (20 ml), leaving a slight excess of the solute. Add a few drops of a saturated sodium acetate solution (about 10 g in 10 ml of water) and then glacial acetic acid slowly until the solution becomes clear and yellow. At this stage the pH is usually 5 but it may range from 2–6. This solution is effective only for about twenty minutes.

The second solution is prepared by partly dissolving the 8-hydroxyquinoline in ethanol as in the first solution. Add a few drops of saturated neutral sodium tartrate solution (about 3 g sodium tartate in 10 ml of water and bring to pH 7 by adding drops of 0.1 *N* HCl). Add concentrated ammonium hydroxide until the solution becomes orange–yellow when the pH is about 10. This solution is effective only for about twenty minutes.

Separate chromatograms are dipped into the solutions and are then placed horizontally on absorbent paper. Two things are worth noting here. First, if the metal hydroxyquinolate is somewhat soluble (as are those of rubidium, lithium and potassium), the fluorescent matter might well pass through the chromatogram to the absorbent paper. If the position of the chromatogram is marked on the latter, the position of the spot may be defined from the absorbent paper. For this reason it is sometimes better to spray on the filtered hydroxyquinoline solutions. Second, the hydroxyquinoline solution sometimes forms fluorescent compounds with the metal ions contained originally in the absorbent paper. These may penetrate to the chromatogram to form irregular patches in contrast to the regular chromatographic spots.

Most of the metal ions produce either fluorescent or coloured hydroxyquinolates (exceptions are Ir^{4+}, TeO_3^{2-}, TeO_4^{-}, SeO_3^{2-}, ReO_4^{-} and Rh^{3+}). After the coloured spots have been marked, the dipped chromatograms are viewed under ultraviolet light (which, of course, must be completely shielded from the human eyes) for observation of the fluorescence. The fluorescent intensity of metal hydroxyquinolates has been studied (STEVENS, 1959) and it has been shown that those which fluoresce most readily are those derived from cations containing the most stable electron groupings. The following scheme gives the metal hydroxyquinolate intensities in order:

$$Al > Y > Sc, Zr, Mg, Sn > Be > Zn > Ag > Li > Na$$

It has been found that sometimes the fluorescence of a hydroxyquinolate is intensified or diminished by exposing the chromatogram to ammonia fumes above a dish or by spraying (POLLARD and MCOMIE, 1953; RITCHIE, 1961). This treatment, which can be carried out under the ultra-violet light, apparently varies the pH, the fluorescence increasing when the optimum pH conditions are approached and vice-versa. Some data on this diagnostic property are given in Tables XXIX and XXX. The author found this technique very suitable for detecting such ions as Mg^{2+}, La^{3+}, Li^{+}, Ca^{2+}, Sr^{2+} and Zr^{4+}.

Special tests are needed for the following ions which are not detected readily by the general methods just described: SeO_3^{2-}, TeO_3^{2-}, TeO_4^{-}, ReO_4^{-}, Ir^{4+} and Rh^{3+}. In addition, more specific tests for Ba^{2+}, Sr^{2+}, K^{+}, Rb^{+} and Cs^{+} are available. Special detecting tests are shown in Table XXXI and some specific detecting reagents are listed in Tables XXXII and XXXV.

TABLE XXIX

DETECTION BY 8-HYDROXYQUINOLINE SOLUTION (pH 2–6)

Ion	*Coloured spot*	*Fluorescence*[1]	*Effect of* NH_3 *spray*[1]
Ag^{+}	faint yellow	f yellow	increased f
Pb^{2+}	yellow	f yellow	
Cd^{2+}	greenish–yellow	ff yellow	
Bi^{3+}	yellow	f yellow (very faint)	
Sn^{2+}	yellow	f yellow	diminished f
Hg^{2+}	yellow	f pale green (faint)	

[1] f = weak fluorescence, ff = strong fluorescence.

TABLE XXIX (continued)

Ion	*Coloured spot*	*Fluorescence*	*Effect of NH_3 spray*
Sb^{3+}	yellow–brown		
As^{3+}	yellow to black		
Fe^{3+}	black		
Cu^{2+}	yellow–brown		
Co^{2+}	brown–yellow		
MoO_4^{2-}	yellow		
Tl^{+}	yellow–brown		
Tl^{3+}	yellow		
Ni^{2+}	green–brown		
Mn^{2+}	brown–black		
Sc^{3+}	faint yellow	ff yellow–green	
Y^{3+}	faint yellow	ff yellow	
La^{3+}	dark yellow	ff green–yellow	increased f
Ga^{3+}	yellow	ff yellow	increased f
In^{3+}	yellow	ff golden yellow	increased f
Zr^{4+}	yellow	f yellow	increased f
Ca^{2+}	very faint yellow	f yellow (faint)	increased f
Sr^{2+}		f yellow (faint)	increased f
Ba^{2+}		f greenish yellow	
Zn^{2+}	greenish yellow	ff green–yellow	
Al^{3+}		ff yellow–green	increased f
Li^{+}		f blue (faint)	increased f
Rb^{+}		f yellow (faint)	diminished f
Cs^{+}		f dull yellow (faint)	
Be^{2+}	yellow	ff yellow	
Mg^{2+}	yellow	ff yellow	increased f
Hf^{4+}	yellow	f yellow	increased f
Th^{4+}	yellow	f orange	increased f
Ge^{4+}	brown (faint)	f yellow (fades)	increased f then fades
Nb^{5+}	yellow	f yellow	increased f
Ta^{5+}		f pale blue	diminished f
Au^{3+}	black		
Pt^{4+}	brown–black (faint)		
Pd^{2+}	yellow (delayed)		
Ru^{+}	black		
Cr^{3+}	dark green		increased spot
CrO_4^{2-}	yellow (faint)		
V^{5+}	brown (faint)		increased spot
WO_4^{2-}	brown	f yellow (very faint)	
Ce^{3+}	brown (faint)		
Ti^{4+}	yellow–brown		reduced spot
UO_2^{2+}	brown		

TABLE XXX

DETECTION BY 8-HYDROXYQUINOLINE SOLUTION (pH 6–10)[1]

Ion	*Coloured spot*	*Fluorescence*[2]	*Effects of* NH_3 *spray*
Sn^{2+}	yellow	f yellow	diminished f
Sb^{3+}	yellow–brown		
Tl^{+}	yellow–brown		
Tl^{3+}	yellow		
Sc^{3+}	yellow (faint)	ff yellow–green	
Y^{3+}	yellow (faint)	ff yellow	
La^{3+}	dark yellow	ff green–yellow	
Ga^{3+}	yellow	ff yellow	
Zr^{4+}	yellow	f yellow	
Zn^{2+}	green–yellow	ff green–yellow	
Li^{+}		f blue (faint)	increased f
Hf^{4+}	yellow	f yellow	increased f
Th^{4+}	yellow	f orange	
Nb^{5+}	yellow	f yellow	diminished f
Ta^{5+}		f pale blue	diminished f
Pd^{2+}	yellow (delayed)		
Ru^{+}	black		
Cr^{3+}	dark green		increased spot
V^{5+}	brown		
WO_4^{2-}	brown	f yellow (very faint)	
UO_2^{2+}	brown		

[1] N.B. Ions tabulated here give more positive results in this solution. For others see Table XXIX.
[2] f = weak fluorescence, ff = strong fluorescence.

The techniques to be adopted when several metal ions are present in the test solution have been discussed already (pp.56–57).

R_F values and chromatographic profiles

The R_F values contained in Tables XXXIIIVnd X XaIX, have been calculated by the usual formula, the distances travelled by the ions being measured either to the centre of the spot or to the top of a comet.

The concept of the *chromatographic profile* has been discussed in the previous chapter (p.56). With few exceptions (notably

TABLE XXXI

SPECIAL IDENTIFICATION TESTS

Test	*Ion*	*Colour of spot*
Dry chromatogram from $(NH_4)_2S_x$ test and then dip into solution of $SnCl_2$ in 5 *N* HCl. Alternatively, spray with the same solution.	SeO_3^{2-}	orange
	TeO_3^{2-}	dark brown to black
	TeO_4^-	brown
	Pt^{4+}	yellow–brown
	Au^{3+}	purple to black
Spray a chromatogram from the 8-hydroxyquinoline test with a fresh solution of sodium rhodizonate (0.5 mg in 2 ml water).	Ba^{2+}	reddish–purple
	Sr^{2+}	brick–red
Spray chromatogram from previous test with a solution of $SnCl_2$ + KCNS + 5 *N* HCl.	ReO_4^-	yellow
	SeO_3^{2-}	yellow
	TeO_3^{2-}	dark brown
	TeO_4^-	brown
	Au^{3+}	brown–black
	Pt^{4+}	golden yellow
Spray unused chromatogram with chlorine water (in open air or in fume cupboard).	Ir^{4+}	faint brown
Spray chromatogram from previous test with a mixture of $SnCl_2$ in 5 *N* HCl to which a little KI has been added. Heat under infra-red lamp.	Rh^{3+}	brown
A chromatogram from the chlorine water test is dried thoroughly to remove the chlorine. The acid character of the chromatogram is neutralized by exposing it to ammonia fumes. It is then sprayed with p-dipicrylamine solution[1] (dissolve 0.2 g dipicrylamine in boiling 0.1 *N* Na_2CO_3 solution; cool and filter). Respray after 5 minutes. After a further 3 minutes it is placed on a glass plate and washed with 0.1 *N* HCl. Faint colour differences are discernable.	K^+	pinkish–red
	Rb^+	bluish–red
	Cs^+	yellowish–red

[1] See RITCHIE (1961).

TABLE XXXII

SPECIFIC DETECTING REAGENTS FOR COMMON METAL IONS[1]

Metal	*Reagent and colour in visible light*	*Reagent and nature of fluorescence in UV-light*
Aluminium	Alizarin (pale)	8-hydroxyquinoline(oxine) (yellow–green)
	Aurin tricarboxylic acid-ammonium salt (pale)	Oxine-kojic acid (green)
	Diphenylcarbazone (red–violet)	Sodium rhodizonate (green)
	Quercetin (green)	
	Quinalizarin (violet)	
Antimony	Alizarin (peach)	Oxine (yellow–green)
	Ammonium polysulphide (yellow)	
	Dithizone (pale)	
	Quercetin (brown)	
	Sodium dithionate (black)	
Arsenic	Ammonium molybdate (blue)	Dimethylaminobenzylidene Rhodamine (rhodanine) (green)
	Alizarin (red–violet)	
	Diphenylcarbazone (pale)	
	Sodium thiosulphate-cupric acetate (brown)	
	Ammonium polysulphide (black)	
Barium	Sodium rhodizonate (red)	Oxine (yellow)
Beryllium	Quercetin (yellow)	Quercetin (brown)
		Oxine (yellow)
Bismuth	Alizarin (red–violet)	Oxine (brown)
	Diphenylcarbazone (red–violet)	
	Sodium dithionate (black)	
	Rubeanic acid (yellow)	
	Dimercaptothiodiazol (bismuthiol) (yellow)	
Cadmium	Ammonium polysulphide (yellow)	Oxine (yellow)
	Titan yellow (red)	
	Diphenylcarbazone (red–violet)	

[1] A modification of data presented by WALDI (unknown date).

TABLE XXXII (continued)

Metal	*Reagent and colour in visible light*	*Reagent and nature of fluorescence in UV-light*
Caesium	p-Dipicrylamine (yellowish–red)	Oxine (yellow)
	Alizarin (red–violet)	Oxine (yellow)
	Diphenylcarbazide (pale)	
	Quinalizarin (violet)	
Cerium	Oxine (brown)	
Chromium	Oxine (dark green)	Oxine (brown)
(Cr^{3+})	Alizarin (red–violet)	
	Quercetin (yellow)	Quercetin (brown)
	Sodium dithionate (blue)	Rhodanine (brown)
	Diphenylcarbazone (violet)	
Cobalt	Ammonium polysulphide (brown–black)	
	Oxine (brown–yellow)	Oxine (brown)
	Diphenylcarbazone (pale)	
	Rubeanic acid (yellow)	
Copper	Ammonium polysulphide (dark brown)	
	Oxine (yellow–brown)	Oxine (yellow)
	Diphenylcarbazone (red–violet)	
	Rubeanic acid (green)	
Gallium	Oxine (yellow)	Oxine (yellow)
Germanium	Oxine (yellow)	Oxine (yellow)
		Phenylfluorone (red)
Gold	Oxine (black)	
	Dithizone (red)	
	Rhodamine B (red)	
Hafnium	Oxine (yellow)	Oxine (yellow)

TABLE XXXII (continued)

Metal	*Reagent and colour in visible light*	*Reagent and nature of fluorescence in UV-light*
Indium	Oxine (yellow) Alizarin (dark red) Morin (faint green)	Oxine (golden–yellow) Morin (green)
Iridium	Hydroquinone (violet) p-Phenylenediamine dihydrochloride (green)	
Iron	Ammonium polysulphide (black) Oxine (black) Alizarin (violet) Sodium rhodizonate (brown)	
Lanthanum	Oxine (dark yellow)	Oxine (yellow–green)
Lead	Ammonium polysulphide (orange to black–brown) Oxine (yellow) Diphenylcarbazone (violet)	 Oxine (yellow)
Lithium	Alizarin (violet) Diphenylcarbazone (green)	Oxine (blue)
Magnesium	Oxine (yellow) Quercetin (brown) Alizarin (red–violet)	Oxine (yellow) Quercetin (yellow)
Manganese	Ammonium polysulphide (brown–black) Oxine (brown–black) Diphenylcarbazide (red–violet)	
Mercury	Ammonium polysulphide (black) Oxine (yellow) Alizarin (red–violet) Dithizone (peach) Sodium rhodizonate (blue grey)	 Oxine (green)

TABLE XXXII (continued)

Metal	*Reagent and colour in visible light*	*Reagent and nature of fluorescence in UV-light*
Molybdenum	Ammonium polysulphide (yellow) oxine (yellow) Potassium ethylxanthogenate (violet)	Morin (faint green)
Nickel	Ammonium polysulphide (brown–black) Oxine (green–black) Dithizone (green) Diphenylcarbazone (violet)	Quercetin (brown)
Niobium	Oxine (yellow) Pyrogallol (yellow)	Oxine (yellow)
Palladium	Oxine (yellow, delayed) Dimethylglyoxime (yellow)	
Platinum	Oxine (dark brown) Benzidine (pale yellow) Rubeanic acid (red–brown)	
Potassium	p-Dipicrylamine (pinkish–red)	Oxine (blue)
Rubidium	p-Dipicrylamine (bluish–red)	Oxine (yellow)
Scandium	Oxine (yellow) Quinalizarin (blue)	Oxine (yellow–green)
Selenium	Stannous chloride (orange)	
Silver	Ammonium polysulphide (black) Oxine (yellow) Dithizone (peach) Sodium rhodizonate (brown)	Oxine (yellow)
Strontium	Sodium rhodizonate (peach)	Oxine (yellow)
Tantalum	Pyrogallol (yellow)	Oxine (blue)
Tellurium	Stannous chloride (brown)	

TABLE XXXII (continued)

Metal	*Reagent and colour in visible light*	*Reagent and nature of fluorescence in UV-light*
Thallium	Ammonium polysulphide (dark-brown) Oxine (yellow–brown) Dithizone (red)	
Thorium	Oxine (yellow) Alizarin (violet)	Oxine (orange)
Tin	Ammonium polysulphide (brown-black) Oxine (yellow)	Oxine (yellow)
Titanium	Oxine (yellow–brown) 1% Chromotropic acid-quercetin (brown)	
Tungsten (wolfram)	Oxine (brown) Rhodamine-B (violet)	Oxine (yellow)
Uranium	Oxine (brown)	
Vanadium	Oxine (brown)	
Zinc	Oxine (green–yellow) Dithizone (pale purple–red)	Oxine (green–yellow)
Zirconium	Oxine (yellow) Alizarin (red–brown)	Oxine (yellow)

Cd^{2+}–Sn^{2+} and V^{5+}–UO_2^{2+}) the chromatographic profile for each ion is characteristic and permits identification of the ion. With experience, one learns to regard the profile as characteristic in itself.

In addition, it is easier to memorize the character of the profiles than the actual R_F values (see Fig. 5). Charts of chromatographic profiles can be made from the data contained in Table XXXIV. Data on the colour of the sulphide, colour of the hydroxyquinolate and the nature of the fluorescence can be shown near each profile. Thus confirmatory data are available for each ion.

TABLE XXXIII

R_F VALUES OF METAL IONS

Ion	R_F Values				Sulphides	Hydroxy-quinolates
	Solvent 1[1]	Solvent 2[1]	Solvent 3[1]	Solvent 4[1]		
Ag^+	0.77+c	1.0 +c	0.13+c	0.59+c	Coloured	Fluorescent and faintly coloured
Pb^{2+}	0.44+c	0.64+c	0.45+c	0.48+c		
Cd^{2+}	0.92	1.0	0.96	0.77–0.83		
Bi^{3+}	0.75	0.85	0.88	0.85		
Sn^{2+}	0.93	0.95–1.0	0.93	0.82		
Hg^{2+}	1.0	1.0	1.0	1.0		
Sb^{3+}	0.88	1.0	0.80+c	0.82	Coloured	Coloured
As^{3+}	0.92	0.92	0.92	0.65		
Fe^{3+}	0.79–0.93	0.65–0.75	0.70–0.73	0.86		
Cu^{2+}	0.53	1.0 and 0.54	0.65	0.56		
Co^{2+}	0.45	0.30–0.38	0.70	0.45–0.55		
MoO_4^{2-}	0.73	0.53+c	0.60	0.83–0.92+c		
Tl^+	0.41+c	1.0 +c	0.41+c	1.0 +c		
Tl^{3+}	1.0	1.0 +c	1.0 +c	1.0		
Ni^{2+}	0.52	0.32	0.63–0.67	0.03		
Mn^{2+}	0.48	0.32	0.70	0.25		
Sc^{3+}	0	0.25	0.75	0	Colourless	Fluorescent and some faintly coloured
Y^{3+}	0	0.18–0.24	0.70–0.75	0		
La^{3+}	0–0.10	0.19	0.65	0.24+c		
Ga^{3+}	1.0	0.65–0.70	0.80	0.63–0.68		
In^{3+}	0.57	0.91–1.0	0.78	0.85–0.90		
Zr^{4+}	0.52–0.60+c	0.12+c	0.52+c	0.48+c		
Ge^{4+}	0.59	0.81	0.47	0.29		
Ca^{2+}	0.40+c	0.19	0.44	0.02		
Sr^{2+}	0.30	0.13	0.40	0	Colourless	Fluorescent and some faintly coloured
Ba^{2+}	0.22	0.06	0.24	0		
Zn^{2+}	0.94	1.0	0.94	0.90		
Al^{3+}	0.42–0.46	0.36	0.75	0.02		
Li^+	0.46	0.40	0.75	0.13		
Rb^+	0.43	0.23	0.35	0.05		
Cs^+	0.46	0.26	0.34	0.03		
K^+	0.41	0.10	0.46	0.05		
Nb^{5+}	0.88	0.26	0.86	0.82–0.94		
Ta^{5+}	1.0	0.90+c	1.0+c	1.0		
Be^{2+}	0.54–0.60	0.60	0.86	0.22		
Mg^{2+}	0.47	0.37	1.0	0.03		
Hf^{4+}	0.53+c	0.11+c	0.50+c	0.46		
Th^{4+}	0	0.15	0.60	0		

TABLE XXXIII (continued)

Ion	*R_F Values*				*Sulphides*	*Hydroxy-quinolates*
	Solvent 1[1]	*Solvent 2*[1]	*Solvent 3*[1]	*Solvent 4*[1]		
Au^{3+}	1.0	0.86–1.0	0.91	0.97		
Pt^{4+}	1.0+c	1.0+c	0.90+c	0.93+c		
Pd^{2+}	0.83	1.0	0.91	0.90		
Ru^{+}	0.30+c	1.0+c	0.70+c	0.82+c		
Cr^{3+}	0.38	0.23	0.73	0.06		
CrO_4^{2-}	1.0+c	0.30+c	0.86+c	1.0+c	Colourless	Coloured
Ce^{3+}	0	0.15	0.54	0		
Ti^{4+}	0.64	0.43	0.76	0.64		
UO_2^{2+}	0.46	0.37–0.43	0.76	0.58–0.64		
V^{5+}	0.47	0.43	0.74	0.64		
WO_4^{2-}	0.70	0	1.0+c	1.0+c		
Ir^{4+}	0.47	0.17	0.82–0.86	0.03		
TeO_3^{2-}	0.86	1.0	0.62	0.71 and 0.15		
TeO_4^{-}	0.68+c	1.0+c	0.62+c	0.71 and 0.15 +c	Colourless	Colourless and non-fluorescent
SeO_3^{2-}	0.72	1.0	0.72	1.0		
ReO_4^{-}	0.82	1.0	0.87	0.89		
Rh^{3+}	0.64	0.43	0.65	0.05		

[1] Solvent 1. Butanol–HCl–HF–water.
Solvent 2. Butanol–HBr–water.
Solvent 3. Ethanol–Methanol–HCl–water.
Solvent 4. Acetone–HCl–HF–water.
(See p.102).

Identification of the metal ions

The metal ions are identified primarily by their R_F values or by their chromatographic profiles. For example, if in the case of a single metal ion, the following chromatographic data were obtained from the chromatograms:

R_F in solvents 1–4 respectively 0.54, 0.53, 0.63 and 0.57.

Colour of sulphide: dark brown.

Colour of hydroxyquinolate: yellow–brown.

Flourescence: absent.

Reference to Table XXXIV will show that for solvent 1, In^{3+},

TABLE XXXIV

NUMERICAL ORDER OF R_F VALUES OF CATIONS[1]

Solvent 1	*Solvent 2*	*Solvent 3*	*Solvent 4*
1.00 Ga^{3+} Ta^{5+}	1.00 In^{3+} Zn^{2+}	1.00 Mg^{2+} Ta^{5+}	1.00 Ta^{5+} Hg^{2+}
1.00 Au^{3+} Pt^{4+}	1.00 Au^{3+} Pt^{4+}	1.00 WO_4^{2-}	1.00 CrO_4^{2-}
1.00 CrO_4^{2-} Hg^{2+}	1.00 Pd^{2+} Ru^{+}	1.00 Hg^{2+} Tl^{3+}	1.00 WO_4^{2-}
1.00 Tl^{3+}	1.00 TeO_3^{2-} ReO_4^-	0.96 Cd^{2+}	1.00 SeO_3^{2-}
0.94 Zn^{2+}	1.00 SeO_3^{2-} TeO_4^-	0.94 Zn^{2+}	1.00 Tl^{+} Tl^{3+}
0.92 Cd^{2+} As^{3+}	1.00 $Ag^{+}$$Cd^{2+}$	0.93 Sn^{2+}	0.97 Au^{3+}
0.91 Sn^{2+}	1.00 Sn^{2+} Hg^{2+}	0.92 As^{3+}	0.93 Pt^{4+}
0.88 Nb^{5+} Sb^{3+}	1.00 Sb^{3+} Cu^{2+}	0.91 Au^{3+} Pd^{2+}	0.90 Zn^{2+} Pd^{2+}
0.86 TeO_3^{2-}	1.00 Tl^{+} Tl^{3+}	0.90 Pt^{4+}	0.89 ReO_4^-
0.83 Pd^{2+}	0.92 As^{3+}	0.88 Bi^{3+}	0.86 Sb^{3+} Fe^{3+}
0.82 ReO_4^-	0.90 Ta^{5+}	0.87 ReO_4^-	0.85 In^{3+} Bi^{3+}
0.79 Fe^{3+}	0.85 Bi^{3+}	0.86 Nb^{5+} Be^{2+}	0.83 MoO_4^{2-}
0.77 Ag^{+}	0.81 Ge^{4+}	0.86 CrO_4^{2-}	0.82 Nb^{5+} Ru^{+}
0.75 Bi^{3+}	0.65 Ga^{3+} Fe^{3+}	0.82 Ir^{4+}	0.82 Sn^{2+}
0.73 MoO_4^{2-}	0.64 Pb^{2+}	0.80 Ga^{3+} Sb^{3+}	0.77 Cd^{2+}
0.72 SeO_3^{2-}	0.60 Be^{2+}	0.78 In^{3+}	0.68 Ga^{3+}
0.70 WO_4^{2-}	0.54 Cu^{2+}	0.76 Ti^{4+} UO_2^{2+}	0.65 As^{3+}
0.68 TeO_4^-	0.53 MoO_4^{2-}	0.75 Sc^{3+} Al^{3+}	0.64 Ti^{4+} V^{5+}
0.64 Ti^{4+} Rh^{3+}	0.43 Ti^{4+} UO_2^{2+}	0.75 Li^{+}	0.64 UO_2^{2+}
0.60 Be^{2+}	0.43 V^{5+} Rh^{3+}	0.74 V^{5+}	0.59 Ag^{+}
0.59 Ge^{4+}	0.40 Li^{+}	0.73 Cr^{3+} Fe^{3+}	0.56 Cu^{2+}
0.57 In^{3+}	0.38 Co^{2+}	0.72 SeO_3^{2-}	0.48 Zr^{4+}
0.53 Hf^{4+} Cu^{2+}	0.37 Mg^{2+}	0.71 TeO_3^{2-}	0.48 Pb^{2+}
0.52 Zr^{4+} Ni^{2+}	0.36 Al^{3+}	0.70 Y^{3+} Ru^{+}	0.46 Hf^{4+}
0.48 Mn^{2+}	0.32 Ni^{2+} Mn^{2+}	0.70 Co^{2+} Mn^{2+}	0.45 Co^{2+}
0.47 Mg^{2+} V^{5+} Ir^{4+}	0.30 CrO_4^{2-}	0.65 La^{3+} Cu^{2+} Rh^{3+}	0.29 Ge^{4+}
0.46 Li^{+} Cs^{+} UO_2^{2+}	0.26 Nb^{5+} Cs^{+}	0.63 Ni^{2+}	0.25 Mn^{2+}
0.45 Co^{2+}	0.25 Sc^{3+}	0.60 Th^{4+} MoO_4^{2-}	0.24 La^{3+}
0.44 Pb^{2+}	0.24 Y^{3+}	0.54 Ce^{3+}	0.22 Be^{2+}
0.43 Rb^{+}	0.23 $Rb^{+}$$Cr^{3+}$	0.52 Zr^{4+}	0.15 TeO_3^{2-}
0.42 Al^{3+} Tl^{+}	0.19 La^{3+} Ca^{2+}	0.50 Hf^{4+}	0.13 Li^{+}
0.41 K^{+}	0.17 Ir^{4+}	0.47 Ge^{4+}	0.06 Cr^{3+}
0.40 Ca^{2+}	0.15 Th^{4+} Ce^{3+}	0.46 K^{+} TeO_4^-	0.05 Rb^{+} K^{+} Rh^{3+}
0.38 Cr^{3+}	0.13 Sr^{2+}	0.45 Pb^{2+}	0.03 Cs^{+} Mg^{2+}
0.30 Sr^{2+} Ru^{+}	0.12 Zr^{4+}	0.44 Ca^{2+}	0.03 Ir^{4+} Ni^{2+}
0.22 Ba^{2+}	0.11 Hf^{4+}	0.42 Tl^{+}	0.02 Ca^{2+} Al^{3+}
0.01 La^{3+}	0.10 K^{+}	0.40 Sr^{2+}	0.00 Sc^{3+} Y^{3+}
0.00 Sc^{3+} Y^{3+}	0.06 Ba^{2+}	0.35 Rb^{+}	0.00 Sr^{2+} Ba^{2+}
0.00 Th^{4+} Ce^{3+}	0.00 WO_4^{2-}	0.34 Cs^{+}	0.00 Th^{3+} Ce^{3+}
		0.24 Ba^{2+}	0.00 TeO_4^-
		0.13 Ag^{+}	

[1] Solvents are those listed in Table XXXIII.

TABLE XXXV

SELECTIVE REAGENTS[1]

Metal	*Reagent*	*Selectivity in p.p.m.*	*Metal*	*Reagent*	*Selectivity in p.p.m.*
Sb	rhodamine B	1	Ni	α-furildioxime	10
As	mercuric chloride	10	Nb	potassium thiocyanate	100
Bi	diethyldithiocarbamate	5	Se	reduction to elemental Se	50
Cr	oxidation to chromate	100	Sn	4,5 dihydroxyfluoresein (gallein)	10
Cu	2,2 biquinoline	10			
Co	2-nitroso-1-naphthol	10	Ti	tiron	150
Ge	phenylfluorone	4	W	potassium thiocyanate	20
Pb	dithizone	20	U	potassium ferrocyanide	4
Mn	oxidation to permanganate	50	V	orthophosphoric acid and sodium tungstate	50
Hg	dithizone	1			
Mo	potassium thiocyanate	1	Zn	dithizone	20

[1] After CANNEY and HAWKINS (1960). Reproduced with the kind permission of the Director, U.S. Geological Survey.

Hf^{4+}, Cu^{2+}, Zr^{4+} and Ni^{2+} on R_F values alone would be possibilities. The coloured sulphide information would eliminate In^{3+}, Hf^{4+} and Zr^{4+}. The colours of the sulphides and hydroxyquinolates of copper and nickel are close enough to be confusing, although, with experience, this should not be so. In solvent 2 however, the R_F value (0.53) eliminates Ni^{2+} but introduces MoO_4^{2-} which is rejected on the following counts:

R_F value in solvent 1 (0.73).

Colour of sulphide and hydroxyquinolate (yellow).

At this stage the identification of Cu^{2+} is positive and is confirmed by the R_F values in solvents 3 and 4.

Results

The paper chromatographic method is so sensitive that impurities, which would normally be neglected or not even observed, show up on the chromatograms. This is particularly true of iron, not only because of its ubiquity, but because of its strong response to the identifying reagents. Hence, great care must be observed when taking the 0.05 g mineral sample, that foreign material is not included.

In general, if a metal constitutes 10% of a mineral, it will show up on the chromatogram. Of course, if the concentration of the solution of the mineral is increased (by dissolving more of the mineral or by partial evaporation of the original solution), the presence of minor constituents might be revealed. In this case, the high concentration of the major constituents might create some difficulties with large dense spots. When the original solution is highly concentrated, metals present by ionic substitution are sometimes detected on the chromatograms and might lead to some confusion.

The use of sensitive specific reagents (see Table XXXII) will eliminate (or at least diminish) the effect of some major constituents and will facilitate the detection of the minor constituent required. The now wellknown spot tests (FEIGL, 1954) will be a great value in these problems, which often will differ from case to case. A great amount of useful research could be achieved by further investigations in this area.

Ore mineral identification — the determinative table

In a hitherto unpublished determinative table (Table XXXVI), the author has compiled data on the metal content of more than one hundred and fifty minerals. A very vague quantitative assessment of the chromatograms is taken into consideration. Based on the size and density of the spots by visual observation, the metals are divided into major and minor constituents. Specific tests for the non-metallic anions are not included. However, the presence of sulphides, carbonates and arsenides is often revealed during the process of dissolution of the mineral if hydrogen sulphide or carbon dioxide or arsine are liberated. In the determinative table constituents detected in this way, or implied from physical examination, are listed in column 4 as "Elements or radicals determined otherwise or inferred". The chemical formulae and approximate metal proportions (column 5) of the minerals permit further the vague quantitative assessment of the chromatograms. The ore minerals have been arranged in

TABLE XXXVI

DETERMINATIVE TABLE

Metals determined on chromatograms			*Elements or radicals determined otherwise or inferred*	*Formulae and approximate metal proportions*	*Mineral*
Metal group	*Other major constituents*	*Minor constituents*			
1	*2*	*3*	*4*	*5*	*6*
			O	Al_2O_3 53:47	Corundum
Al		Fe	O,OH	$Al_2O_3.2H_2O$ 37:73	Bauxite, etc.
		K	O,SO_4	$K_2Al_6(OH)_{12}(SO_4)_4$ 8:20	Alunite
			Na,F	$AlNa_3F_6$ 13:32	Cryolite
				Sb 100	Antimony(native)
			S	Sb_2S_3 72:28	Stibnite
	Ag			Ag_3Sb 73:27	Dyscrasite
	Ag		S	Ag_3SbS_3 59:22:18	Pyrargyrite
	Ag		S	$(Ag,Cu)_{16}Sb_2S_{11}$ 69:8:9:15	Polybasite
	Pb	Ag	S	$PbAgSb_3S_6$ 24:12:41:22	Andorite
	Ni			NiSb 32:67	Briethauptite
Sb	Ni		S	NiSbS 28:57:14	Ullmannite
	Pb		S	$Pb_6Sb_8S_{17}$ 33:45:23	Zinkenite
	Pb	As	S	$Pb_5(Sb,As)_2S_8$ 70:8:5:17	Geocronite
	Pb		S	$Pb_5Sb_4S_{11}$ 51:29:20	Boulangerite
	Pb	Cu	S	$PbCuSbS_3$ 43:13:25:20	Bournonite
	Pb	Fe	S	$Pb_4FeSb_6S_{11}$ 40:3:34:21	Jamesonite
	Cu		S	Cu_3SbS_4 43:28:29	Famatimite
	Fe		S	$FeSb_2S_4$ 34:31:35	Berthierite
	Cu		S	$(Cu,Fe)_{12}Sb_4S_{13}$ 45:1:29:24	Tetrahedrite
				As 100	Arsenic (native)
			S	AsS 70:30	Realgar
			S	As_2S_3 61:39	Orpiment
	Fe		S	FeAsS 34:46:20	Arsenopyrite
	Fe			$FeAs_2$ 27:73	Loellingite
	Co	Fe		$(Co,Fe)As_2$ 19:9:70	Safflorite
	Co		S	CoAsS 35:45:19	Cobaltite
	Cu			Cu_3As 72:28	Domeykite
As	Cu	Fe	S	$(Cu,Fe)_{12}As_4S_{13}$ 48:5:19:28	Tennantite
	Cu		S	$Cu_3(As,Sb)S_4$ 48:18:1:32	Enargite
	Ni			$NiAs_2$ 28 :67	Rammelsbergite
	Co	Ni		$(Co,Ni)As_3$ 20:5:75	Skutterudite
	Co,Fe		S	CoFeAsS 11:23:46:20	Glaucodot
	Pb		S	$Pb_2As_2S_5$ 57:21:22	Dufrenoysite

TABLE XXXVI (continued)

1	2	3	4	5	6
As. (cont.)	Pb	Sb	S	$Pb_5(Sb,As)_2S_8$ 70:8:5:17	Geocronite
	Pb	Cu	S	$PbCuAsS_3$ 47:14:17:22	Seligmannite
	Pb			$Pb_5(AsO_4,PO_4)_3Cl$ 70:8:5:17	Mimetite
	Ag		S	$3Ag_2S.As_2S_3$ 65:15:19(S)	Proustite
	Pt			$PtAs_2$ 56:42	Sperrylite
Ba			SO_4	$BaSO_4$ 59:41	Barite
			CO_3	$BaCo_3$ 70:30	Witherite
Be	Al			$Be_3Al_2(SiO_3)_6$ 5:10:31:54	Beryl
Bi				Bi 100	Bismuth (native)
			O	Bi_2O_3 90:10	Bismite
			CO_3	$(BiO)_2.CO_3.H_2O$ 90	Bismuthite
			S	Bi_2S_3 81:19	Bismuthinite
	Pb		S	$PbS.Bi_2S_3$ 28:55:17(S)	Galenobismutite
	Te			Bi_2Te_3 52:48	Tellurobismuthite
	Te		S	Bi_2Te_2S 59:36:5	Tetradymite
Cd			S	CdS 78:22	Greenockite
Ca			CO_3	$CaCO_3$ 40:60	Calcite
			CO_3	$CaCO_3$ 40:60	Aragonite
	Mg		CO_3	$(Ca,Mg)CO_3$ 21:13:56	Dolomite
	Mn		CO_3	$(Ca,Mn)CO_3$ 20:20:60	Manganocalcite
	Fe		CO_3	$(Ca,Fe)CO_3$ 19:19:62	Ankerite
			F	CaF_2 51:49	Fluorite
			SO_4	$CaSO_4$ 30:70	Anhydrite
			H_2O,SO_4	$CaSO_4.2H_2O$ 23:57:20	Gypsum
	U			$Ca(UO_2)_2P_2O_8.8H_2O$ 4:48	Autunite
Cr	Fe		O	$FeCr_2O_4$ 25:46:29	Chromite
	Pb		O	$PbCrO_4$ 64:14:22	Crocoite
Co			S	Co_3S 49:42	Linnaeite
	As		S	CoAsS 29:45:19	Cobaltite
	Fe,As		S	CoFeAsS 11:23:46:20	Glaucodot
	Fe,As			$CoFeAs_2$ 19:9:70	Safflorite
	As	Ni		$(Co,Ni)As_3$ 20:5:75	Skutterudite
	Ni		S	$(Co,Ni)_3S_4$ 26:31:43	Siegenite
	As		H_2O	$Co_3As_2O_8.8H_2O$ 26:25:25:24	Erythrite

TABLE XXXVI (continued)

1	*2*	*3*	*4*	*5*	*6*
				Cu 100	Copper (native)
			O	CuO 80:20	Tenorite
			O	Cu_2O 89:11	Cuprite
			CO_3,OH	$CuCO_3.Cu(OH)_2$ 57:35:8	Malachite
			CO_3,OH	$2CuCO_3.Cu(OH)_2$ 54:41:5	Azurite
			S	CuS 66:34	Covellite
			S	Cu_2S 80:20	Chalcocite
			S	$Cu_{2-x}S$ 79:21	Digenite
	Fe		S	$CuFeS_2$ 34:31:35	Chalcopyrite
		Fe	S	Cu_5FeS_4 63:11:26	Bornite
Cu	Fe,Sn		S	Cu_2FeSnS 30:13:28:30	Stannite
	As			Cu_3As 72:28	Domeykite
	As	Fe	S	$(Cu,Fe)_{12}As_4S_{13}$ 48:5:19:28	Tennantite
	As		S	$Cu_3(As,Sb)S_4$ 48:18:1:32	Enargite
	Sb		S	$(Cu,Fe)Sb_4S_{13}$ 45:1:29:24	Tetrahedrite
	Sb		S	Cu_3SbS_4 43:28:29	Famatimite
	Pb,Sb		S	$PbCuSbS_3$ 42:13:25:20	Bournonite
	Pb,As		S	$PbCuAsS_3$ 47:14:17:22	Seligmannite
	Pb		SO_4,OH	$PbCu(SO_4)(OH)_2$ 56:19:15:10	Linarite
	Ag		S	Ag CuS 53:31:16	Stromeyerite
	Se			Cu_2Se 62:38	Berzelianite
	U		PO_4,H_2O	$Cu(UO_2)_2P_2O_8.8H_2O$6:47:6:21:20	Torbernite
				Au up to 100	Gold (native)
	Ag			AuAg 74:21	Electrum
	Te			$AuTe_2$ 44:56	Calaverite
Au	Te			$AuTe_2$ 44:56	Krennerite
	Ag,Te			Ag_3AuTe_2 42:25:33	Petzite
	Pb,Te	Sb	S	$Pb_5Au(Te,Sb)_4S_8$ 57:7:18:7:10	Nagyagite
Ir			Os	IrOs 58:35	Iridosmine
				Fe 100	Iron (native)
			O	Fe_3O_4 72:28	Magnetite
			O	Fe_2O_3 70:30	Hematite
			O,H_2O	$Fe_2O_3.2H_2O$ 50:40:10	Goethite
	Cr		O	$FeCrO_4$ 26:45:29	Chromite
Fe	Ti		O	$FeTiO_3$ 35:32:32	Ilmenite
			CO_3	$FeCO_3$ 48:52	Siderite
	Fe	Mn	WO_4	$(Fe,Mn)WO_4$ 19:5(max.):76	Huebnerite
	Mn		O	$(Mn,Fe)_2O_3$ 32:33:25	Bixbyite
			S	FeS_2 46:54	Pyrite
			S	$Fe_{x-1}S$ 64:36	Pyrrhotite

TABLE XXXVI (continued)

1	*2*	*3*	*4*	*5*	*6*
Fe			S	FeS_2 46:54	Marcasite
(cont.)	As		S	FeAsS 34:46:20	Arsenopyrite
	As	Co		$(Fe,Co)As_2$ 23:4:71	Loellingite
	Co,As			$(Co,Fe)As_2$ 19:9:70	Safflorite
	Cu		S	$CuFeS_2$ 34:31:35	Chalcopyrite
	Cu		S	Cu_5FeS_4 64:11:25	Bornite
	Ni		S	$(Ni,Fe)S_2$ 23(max.):24:54	Bravoite
	Ni		S	$(Fe,Ni)_9S_8$ 33:34:33	Pentlandite
	Co,As		S	(Co,Fe)AsS 11:23:46:20	Glaucodot
	Nb	Ta,Mn		$(Fe,Mn)(Nb,Ta)_2O_6$ 13:4:50:4:29	Columbite
	Ta	Nb,Mn		$(Fe,Mn)(Ta,Nb)_2O_6$ 9:5:53:14:19	Tantalite
	Cu,Sn		S	Cu_2FeSnS 30:13:29:30	Stannite
	Ca		CO_3	$(Ca,Fe)(CO_3)_2$ 28:29:43	Ankerite
				Pb 100(max.)	Lead (native)
			S	PbS 87:13	Galena
	Cu		SO_4,OH	$PbCu(SO_4)(OH)_2$ 56:19:15:10	Linarite
	Sb		S	$Pb_5Sb_4S_3$ 51:29:19	Boulangerite
	Sb	Fe	S	$Pb_4FeSb_6S_{14}$ 60:3:35:22	Jamesonite
	Sb	Cu	S	$PbCuSbS_3$ 42:13:25:20	Bournonite
	Sb		S	$Pb_6Sb_{14}S_{27}$ 33:45:23	Zinkenite
	Sb		S	$Pb_5Sb_8S_{17}$ 41:38:21	Plagionite
	Sb		S	$Pb_{13}Sb_7S_{24}$ 61:18:17	Meneghinite
	Sb	Ag	S	$Pb_5Ag_2Sb_6S_{15}$ 42:8:29:20	Owyheeite
	Sb	Ag	S	$PbAgSb_3S_6$ 24:12:41:22	Andorite
	Mo		O	$PbMoO_4$ 56:26:18	Wulfenite
		Sb,As	S	$Pb_5(Sb,As)_2S_8$ 80:8:5:17	Geocronite
	As	Cu	S	$PbCuAsS_3$ 47:14:17:22	Seligmannite
Pb			CO_3	$PbCO_3$ 77:23	Cerussite
			SO_4	$PbSO_4$ 50:50	Anglesite
	Te	Au,Sb	S	$Pb_5Au(Te,Sb)_4S_8$ 57:7:18:7:10	Nagyagite
	As		S	$Pb_2As_2S_5$ 57:21:22	Dufrenoysite
	Mn		O	$MnPbMn_6O_{14}$ 27(Pb) 52(Mn)	Coronadite
		V		$Pb_5(VO_4)_3Cl$ 73:11(V)	Vanadinite
	WO_4			$PbWO_4$ 45:55	Stolzite
	WO_4			$PbWO_4$ 45:55	Raspite
		As	PO_4	$Pb_5(PO_4,AsO_4)_3Cl$ 81:16:trace:3	Pyromorphite
		As	PO_4	$Pb_5(AsO_4,PO_4)_3Cl$ 75:23:trace:2	Mimetite
	Te			PbTe 62:38	Altaite
	Se			PbSe 72:28	Clausthalite
	Bi		S	$PbS.Bi_2S_3$ 28:55:17(S)	Galenobismutite
	CrO_4			$PbCrO_4$ 64:16:20	Crocoite

TABLE XXXVI (continued)

1	*2*	*3*	*4*	*5*	*6*
			CO_3	$MgCO_3$ 27:73	Magnesite
Mg	Ca		CO_3	$(Ca,Mg)CO_3$ 31:13:56	Dolomite
	Fe		CO_3	$(Mg,Fe)CO_3$ 14:24:62	Breunnerite
				$H_2(Ni,Mg)SiO_4.H_2O$ (variable)	Garnierite
			O	MnO_2 63:37	Pyrolusite
			Si,O	$(Mn,Si)_2O_3$ 62:38	Braunite
	Ba		O,OH	$BaMn^{2+}Mn^{4+}O_{16}(OH)_4$ 16:7:73	Psilomelane
				$BaO.MnO_2.Fe_2O_3$ 9:50:4:37	Wad
				$MnMn_2O_4$ 93:7	Hausmannite
Mn	Fe,WO_4			$(Fe,Mn)WO_4$ 19:5(max.):76	Ferberite
	WO_4	Fe		$(Mn,Fe)WO_4$ 19:5(max.):76	Huebnerite
			CO_3	$MnCO_3$ 48:52	Rhodochrosite
			SiO_3	$MnSiO_3$ 42:58	Rhodonite
	Ca		CO_3	$(Mn,Ca)CO_3$ 20:20:60	Manganocalcite
	Pb		O	$MnPbMn_6O_{14}$ 27(Pb)52(Mn)	Coronadite
	Fe		O	$(Mn,Fe)_2O_3$ 32:33:25	Bixbyite
			S	HgS 86:14	Cinnabar
Hg		Se	S	Hg(Se,S) 81:6:10	Onofrite
	Se			HgSe 72:28	Tiemannite
	Te			HgTe 62:38	Coloradoite
			S	MoS_2 60:40	Molybdenite
Mo			O	MoO_3 67:33	Molybdite
	Pb		O	$PbMoO_4$ 56:28:18	Wulfenite
		Fe,Mn,Ta	O	$(Fe,Mn)(Nb,Ta)_2O_6$ 10(max.)10(max.)60:10:20	Columbite
Nb					
	Ta	Fe,Mn	O	$(Fe,Mn)(Ta,Nb)_2O_6$ 12:7:50:15:16	Tantalite
			S	NiS 64:26	Millerite
	Fe		S	(Fe,Ni)S 29:36:33	Pentlandite
	As			NiAs 44:56	Niccolite
		Mg		$H_2(Ni,Mg)SiO_4.H_2O$ (variable)	Garnierite
Ni	Co,As			$(Co,Ni)As_3$ 20:5:75	Skutterudite
	Sb		S	NiSbS 28:57:14	Ullmannite
	Sb			NiSb 32:67	Breithauptite
	Co		S	$(Co,Ni)_3S$ 26:31:43	Siegenite
	Fe		S	$(Ni,Fe)S_2$ 23(max.):24:54	Bravoite
	As			$NiAs_2$ 28:67	Rammelsbergite

TABLE XXXVI (continued)

1	*2*	*3*	*4*	*5*	*6*
Pt				Pt up to 100	Platinum
	As			$PtAs_2$ 53:41	Sperrylite
	Al		OH,SO_4	$K_2Al_6(OH)_{12}(SO_4)_4$ 8(K):20(Al)	Alunite
K	U	V	O,OH	$K_2(UO_2)_2(VO_4)3H_2O$ 8(K):50(U):11(V)	Carnotite
	Pb			PbSe 72:28	Clausthalite
Se	Cu			Cu_2Se 62:38	Berzelianite
	Hg			HgSe 72:28	Tiemannite
	Hg		S	Hg(Se,S) 81:6:10	Onofrite
				Ag 100(max.)	Silver (native)
	Au			Ag,Au 21:74	Electrum
	Sb			Ag_3Sb 73:27	Dyscrasite
			S	Ag_2S 87:13	Argentite
	Sb		S	$(Ag,Cu)_{16}Sb_2S_{11}$ 69:8:9:15	Polybasite
	Te			Ag_2Te 63:36	Hessite
	Te,Au			$(Au,Ag)Te_2$ 24:13:63	Sylvanite
Ag	As		S	$3Ag_2S.As_2S_3$ 65:15:19(S)	Proustite
	Sb		S	$3Ag_2S.Sb_2S_3$ 59:22:18	Pyrargyrite
	Cu		S	$(Ag,Cu)_2S$ 53:31:16	Stromeyerite
			Cl	AgCl 75:25	Cerargyrite
			Br,Cl	Ag(Br,Cl) 65:23:12	Embolite
	Pb,Sb		S	$PbAgSb_3S_6$ 24:12:41:22	Andorite
	Pb,Sb		S	$Pb_5Ag_2Sb_6S_{15}$ 42:8:29:20	Owyheeite
	Au,Te			Ag_3AuTe_2 42:25:33	Petzite
Sr			CO_3	$SrCO_3$ 56:44	Strontianite
			SO_4	$SrSO_4$ 47:53	Celestite
		Nb,Fe,Mn	O	$(Fe,Mn)(Ta,Nb)_2O_6$ 12:7:50:15:16	Tantalite
Ta	Nb	Fe,Mn	O	$(Fe,Mn)(Nb,Ta)_2O_6$ 10(max.):10(max.):60:10:20	Columbite
	Pb			PbTe 62:38	Altaite
	Au			$AuTe_2$ 44:56	Calaverite
	Ag			Ag_2Te 63:36	Hessite
	Hg			HgTe 63:38	Coloradoite
Te	Au			$AuTe_2$ 44:56	Krennerite
	Ag,Au			Ag_3AuTe_2 42:25:33	Petzite
	Au,Ag			$(Au,Ag)Te_2$ 24:13:63	Sylvanite
	Bi			Bi_2Te_3 52:48	Tellurobismuthite
	Bi		S	Bi_2Te_2S 59:36:5	Tetradymite

TABLE XXXVI (continued)

1	*2*	*3*	*4*	*5*	*6*
				$ThSiO_4$ 72:28	Thorite
Th				ThO_2 80:20	Thorianite
			PO_4	$(Ca,La,Y,Th)_6PO_4$ 9% (Th)	Monazite
Sn			O	SnO_2 79:21	Cassiterite
	Cu,Fe		S	Cu_2FeSnS 30:13:28:30	Stannite
Ti			O	TiO_2 61:39	Rutile
	Fe		O	$FeTiO_3$ 37:29:34	Ilmenite
	Ca		O	$CaWO_4$ 14:63:23	Scheelite
	Pb		O	$PbWO_4$ 45:55	Stolzite
W	Pb		O	$PbWO_4$ 45:55	Raspite
	Fe	Mn	O	$(Fe,Mn)WO_4$ 19:5(max.):76	Ferberite
	Mn	Fe	O	$(Mn,Fe)WO_4$ 19:5(max.):76	Heubnerite
			O	UO_2 62:38	Uraninite
U		Cu	P_2O_5,O,H_2O	$Cu(UO_2)_2P_2O_8.8H_2O$ 6:47:6:21:20	Torbernite
		Ca	P_2O_5,O,H_2O	$Ca(UO_2)_2P_2O_8.10H_2O$ 5:48:6:21:20	Autunite
	V	K	O,H_2O	$K_2(UO_2)_2(VO_4)_2.3H_2O$ 8:49:16:22:5	Carnotite
V	Pb		O	$Pb_4(Pb,Cl)(VO_4)_3$ 72:2:11:15	Vanadinite
	U	K	O,H_2O	$K_2(UO_2)_2(VO_4)_2.3H_2O$ 8:49:16:22:5	Carnotite
			S	ZnS 67:33	Sphalerite
			S	ZnS 67:33	Wurtzite
Zn			CO_3	$ZnCO_3$ 52:48	Smithsonite
			SiO_2	Zn_2SiO_4 58:42	Willemite
			SiO_2,H_2O	H_2ZnSiO_5 54% (Zn)	Calamine
Zr			SiO_2	$ZrSiO_4$ 57% (Zr)	Zircon

metal groups (column 1) according to the major constituents. Thus the mineral dyscrasite appears in both the antimony and the silver groups while stannite appears in the copper, iron and tin groups.

Of course, when two or more ore minerals have the same major metal constituents (for example zinkenite, $Pb_6Sb_{14}S_{27}$ and plagionite, $Pb_5Sb_8S_{17}$), they cannot be distinguished by qualitative chemical methods. In this respect, the chromatographic method has the same limitation as other chemical methods.

PLATE I

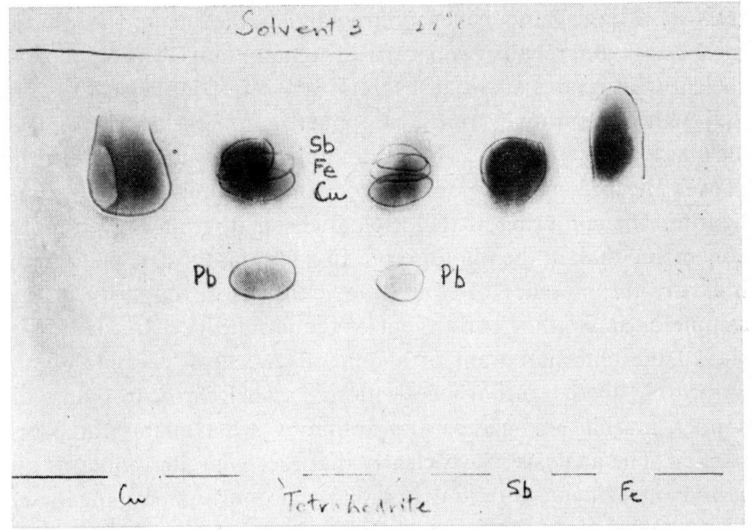

Chromatographic analysis of tetrahedrite and of solutions containing copper, antimony and iron.

PLATE II

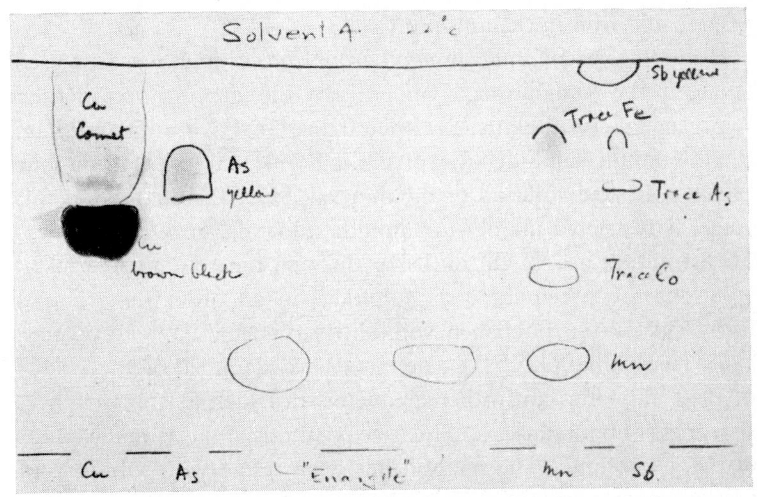

Chromatographic analysis of a mineral labelled "enargite" and of copper, arsenic, manganese and antimony solutions.

Results obtained in both laboratory investigations and in field tests at a base camp have confirmed that the method is quick, inexpensive and reliable. A parallel chromatogram (Plate I), depicts the chromatograms of two test solutions of tetrahedrite [$(Cu,Fe)Sb_4S_{13}$] with comparative "runs" of copper, antimony and iron. The chromatogram (treated with ammonium polysulphide), clearly proves the presence of copper, iron and antimony in the tetrahedrite. The colouring of the original chromatogram is even more convincing than in the photograph. In addition, the chromatogram indicates the presence of some lead, either as an impurity in the sample or as a minor constituent of the tetrahedrite. In Plate II is shown the chromatogram of a mineral, labelled "enargite" [$Cu_3(As, Sb)S_4$] in the author's collection. Parallel chromatograms of copper, arsenic, manganese and antimony are available for comparison. The analysis shows clearly that the "enargite" contains no copper, arsenic nor antimony but a large proportion of manganese. The mineral, in fact, is manganite. Reference to the physical properties of the two minerals will show how closely they resemble each other. In addition, the chromatogram shows the presence of some cobalt and iron as impurities in the manganese and some arsenic and iron in the antimony.

Recently, *precipitation chromatography* on columns has been used to aid in the recognition of minerals by the identification of their major metal constituents (SPAIN et al., 1962). The stationary phase is ammonium sulphide (the precipitant) in agar gel. The mobile phase is an acid solution of the mineral. Since the method, at this stage, is restricted mainly to sulphide minerals, no great problem of dissolution arises. On diffusing through the column, the metal solutions are precipitated as sulphides in an order that follows fairly closely the published solubilities (GREEN, 1959; HODGMAN, 1958) (see Table XXXVII). The metals are identified by the colour of their sulphides and other characteristics such as transparency, sharpness of boundaries and relative positions of the sulphide bands in the gel column. The method has been tested with solutions of known metal content and with mineral solutions which were analysed spectrographically.

TABLE XXXVII

METAL SULPHIDE PRECIPITATION IN AGAR GEL[1]

Metal and solubility of sulphide[2] *g/l*	*Metal in agar gel in descending order*	*Effects in solution above gel meniscus*	*Major sulphide band in gel*
Sn^{4+}0.00002	Sn^{4+}	Colourless then white gel of $Sn(OH)_4$ at 48 hours[3]	Transparent yellow SnS_2
Hg^{2+} insol.	Hg^{2+}	Colourless solution	Opaque black HgS
Cu^{2+}0.001	Cu^{2+}	Green solution $CuCl_4^{2-}$	Opaque greenish-black to thin brown at bottom CuS
Pb^{2+}0.124	Pb^{2+}	Colourless solution, white crystals of $PbCl_2$[3]	Medium grey transparent PbS Liesegang rings at high concentration
Bi^{3+}0.00018	Bi^{3+}	Colourless solution, white BiOCl below meniscus[3]	Opaque dusky brown with ragged lower boundary, Bi_2S_3
Cd^{2+}0.0013	Cd^{2+}	Colourless solution	Opaque yellow, greenish yellow at top and bottom of band, CdS
Sb^{3+}0.00175	$Sb^{3+(5+)}$	Colourless solution, white SbOCl below meniscus[3]	Opaque orange with diffuse lower boundary Sb_2S_3
As^{3+}0.0005	$As^{3+(5+)}$	Colourless solution	Transparent yellow, sharp lower boundary, As_2S_3.
Zn^{2+}0.0069	Zn^{2+}	Colourless solution	Opaque white ZnS
Mo^{6+} insol.	Mo^{6+}	Colourless solution, changes to dark blue in 48 hours	Transparent moderate brown with diffuse upper and lower boundaries, MoS_3

[1] A modification of a table by SPAIN et al. (1962).
[2] HODGMAN (1958).
[3] Indicates an observation not always observed.

TABLE XXXVII (continued)

Metal and solubility of sulphide g/l[2]	*Metal in agar gel in descending order*	*Effects in solution above gel meniscus*	*Major sulphide band in gel*
Te^{4+} insol.	Te^{4+}	Colourless solution	Opaque, sharp moderate brown, TeS_2
Co^{2+}0.0038	Co^{2+}	Pink solution	Transparent brownish-black with grey Liesegang, CoS
Ni^{2+}0.0036	Ni^{2+}	Slightly greenish solution	Brownish-black, transparent NiS
Fe^{2+}0.0062	$Fe^{2+(3+)}$	Greenish-yellow solution	Transparent greenish-black, diffuse upper and lower boundary, FeS

Preparation of gel columns

Gel columns, if stored in an atmosphere of propane or natural gas, last up to six days. It is convenient to make up batches of one hundred or more at a time. Soft glass tubing (6 mm bore) is cut into 24 cm lengths, constricted by drawing out at the centre and then cut into halves. The wide ends are fire polished. Then proceed as follows.

Place 4 g agar agar (shredded) in a 1,000 ml beaker and add enough water to cover it. Let stand about one hour until the shreds are swollen and the soluble pigment is leached away. Pour off excess water. Add the gel to another beaker containing a boiling solution of sodium acetate trihydrate, N.F. grade, (54 g in 200 ml water). Keep boiling for about 5 minutes or until only occasional shreds of agar remain undissolved. Filter while hot through a glass wool pad into a storage flask. The batch is enough for about 120 columns and will keep indefinitely in a stoppered flask. The glass tubes are filled as follows: place the hot stock solution (re-warmed if necessary) into a test tube (22 × 100 mm), add 3 drops ammonium sulphide (A.R. grade) and stir. Place the glass tube (constricted end downwards) into the gel and wait until it rises about two-thirds of the tube. The tube is then withdrawn (with a finger pressed tightly over the wide end) and cooled in ice or cold water until the gel sets. If bubbles form at the top of the gel in the tube, and cannot be dislodged by tapping, the tubes should be discarded. The filled tubes will not keep even 24 hours unless stored in an atmosphere of propane or natural gas.

Preparation of samples for analysis

Place the finely ground minerals (0.5–1.0 g) in a 20 ml porcelain casserole and add concentrated HCl (5 drops) and concentrated HNO_3 (5 drops). Evaporate to dryness and then add 6 *M* HCl (1 ml). Transfer to a test tube. The quantity of sample necessary is determined by trial. Take 2 drops of the mineral solution and add 6 *M* HCl (8 drops). Place 3 drops of this solution on top of the gel column. To avoid the formation of bubbles, place the tip of an eye-dropper near the top of the gel at the centre and apply the solution drop by drop. Bubbles may be removed by tapping or drawn out by the tip of the eye-dropper. Stand the column upright for 48 hours in the same test tube to which the sample solutions were transferred. It is stated to be an advantage to allow diffusion of some of the sample solution upwards from the bottom of the column. Then the colours of bands and other characteristics are observed and interpreted (see Table XXXVII).

Thus precipitation chromatography has been shown to provide an alternative method to paper chromatography for the identification of metal ions in minerals. Each method seems capable of further refinements and so, perhaps, it is too soon to favour one or the other.

ANALYSIS OF MAGNESIAN LIMESTONES

A method which makes use of circular paper chromatography (HJELLE, 1961) provides a rapid and easy means of determining the approximate calcium and magnesium content of magnesian limestones. When hydrochloric acid solutions of calcium and magnesium salts are mixed with suitable indicators, and then chromatographed radially, sorption figures (or rings) are developed, the position of the bands being more or less dependent on the Ca–Mg proportion in the solution. Blue ink (for example Parker Quink) is a suitable indicator. High magnesium content gives a pronounced outer ring whereas high calcium content gives a pronounced inner ring. Low content in each case gives faint colour in the same areas. Absence of one or the other is indicated by a general absence of sorbed dye in the respective areas.

The sorption, and hence the appearance of the rings, depends especially on:

(1) amount and quality of added indicator,
(2) amount of solution added,
(3) pH of solution,
(4) type of absorbent (filter paper),
(5) concentration of Ca–Mg in solution.

If the first four factors are kept constant, the rings produced from test solutions may be interpreted by comparison with rings prepared from solutions of known Ca–Mg content.

The procedure recommended is as follows: Prepare standard solutions of $CaCl_2$ and $MgCl_2$ by dissolving to saturation in each case the carbonate in small amounts of 10% HCl (2.9 *N*). Artificial mixtures with the Ca–Mg content desired are prepared by mixing the two chloride solutions in the necessary proportions.

Add indicator (blue ink) to the mixture in the proportion 4 : 10 by volume. Apply 3 drops of the test solution to the centre of a horizontally placed filter paper. After the solution has spread as far as possible, dry at 20° C.

Tests with "unknown" mixtures of different proportions revealed maximum errors of 10% in the percentage of CaO and average errors of 2.5%. Tests with actual limestones and dolomites revealed a maximum error of 13.2% in the percentage of CaO and an average error of 6.2%.

The method is claimed to be a simple and quick one for the rough classification of calcium-magnesium carbonates.

COPPER, GOLD AND SILVER ASSAYS

Precipitation chromatography on columns or in "batches" has been used for the separation of copper from lead (ZIEGLER, 1961), silver from lead (ZIEGLER and GIESELER, 1962) and gold from transitional metals such as iron, copper and cobalt (ZIEGLER and MATSCHKE, 1962). The initial precipitating materials are cadmium sulphide and cadmium sulphide impregnated cellulose (of M. Ziegler and manufactured by Macherey, Nagel and Co., Düren, Germany). Since, according to the authors, cadmium sulphide and lead sulphide have the greatest solubilities of the acid-insoluble metal sulphides, all other metals of the group will be precipitated as sulphides in preference to cadmium or lead according to the equations

$$CdS + M^{2+} \rightarrow MS + Cd^{2+}$$
$$PbS + M^{2+} \rightarrow MS + Pb^{2+}$$

Under conditions of high concentration of lead, some lead sulphide is precipitated. However, this may be avoided by diluting the sample solution or by adding more cadmium in the form of cadmium nitrate. The lead solution is not contaminated with organic substances but only with a quantity of cadmium equivalent to the metal sulphides precipitated. Thus, if a sample solution containing copper and lead is passed through a column of the cadmium sulphide cellulose, the copper is retained on the column and the lead (with some released cadmium) passes out of the column. Likewise, in a batch process the copper is precipitated on the treated cellulose while lead and cadmium remain in solution. Similarly, solutions containing silver and lead can be separated. To separate gold from other metals (including bismuth, copper and platinum), it is better to convert the cadmium sulphide cellulose to silver sulphide cellulose, which in turn is replaced by the gold alone on the column.

In each case the precipitated sulphides with or without the cellulose are ashed and the desired determination of the metal is made by using gravimetric, colorimetric, X-ray fluorescent or flame photometric methods of analysis.

The column is packed with great care to achieve uniformity. Small quantities in the form of a slurry are added at ten minute intervals with suction being applied at the lower end of the equipment. The suction is adjusted so that a flow rate of 200 ml per hour is obtained. Each thin layer is pressed with a ramrod before the next quantity of slurry is added. The lower end of the column tube is ground flat and rests on a filter paper pad and is tightly fitted to a sintered glass crucible or a perforated plate. The suction pump is connected to the crucible or plate.

Copper from lead

Dissolve the copper–lead sample as nitrates and make up to 1% HNO_3 (100 ml) in a beaker. Add 800 mg CdS and stir for about 15 minutes with a magnetic stirrer. Filter on a sintered glass crucible (type 1G4), the filtrate being retained. Wash precipitate with 5% NaOH (10 ml) and then with water. Ash the copper sulphide in the crucible and then place in 150 ml beaker with 10–20 ml water. Add conc. HNO_3 (2–4 ml) and warm. The sulphide dissolves rapidly leaving some free sulphur. Make up the copper solution to a known volume (50 ml). Estimation is made photometrically for solutions containing 3–30 mg Cu, as Cu(II)-amino-acetate at 537 mμ. For amounts ranging from 0.3–3 mg estimation is made photometrically with violuric acid at 415 mμ. The reagent is added before the solution is made up to its known volume. Of the metals in the filtrate, lead can be determined with E.D.T.A. (FLASCHKA, 1959), cobalt photometrically with rhodanine (dimethylaminobenzylidene rhodamine) and iron as Fe_2O_3. The efficacy of the method is shown by the results taken from the many in the original paper (see Table XXXVIII).

Silver from lead (ZIEGLER and GIESELER, 1962)

MN-cellulosepulver 2100 CdS-8% is packed to a height of 3 cm into a tube 15 cm × 1.8 cm. The powder is washed with 0.01 N HNO_3. The test solution should contain up to 20 mg of silver in a volume up to 10 litres. It is acidified with HNO_3 and the pH is adjusted within the range 1.4–2.4 by adding either acid or alkali. The flow rate should be 5 ml per minute. After the Ag_2S precipitates (as a black zone), the column is washed with 0.01 N HNO_3 (100–150 ml).

TABLE XXXVIII

COPPER–LEAD ESTIMATIONS IN PREPARED SAMPLES

Given (mg)	Found (mg)	Error %
Copper		
25.5	25.5	± 0
9.6	9.65	± 0.5
1.03	1.03	± 0
Lead		
55.9	55.9	± 0
978.5	982	± 0.4
1955	1958	± 0.15

The sulphide cellulose is then extruded from the column by air pressure applied from below. The Ag_2S is dissolved by repeated warming in 4 volumes of 7 *N* HNO_3 (20 ml). Each volume is then cooled and filtered on a glass filter (type 1G4). Finally, the filter is washed with aqueous 10% NH_4OH (10 ml). The combined filtrates are reduced in volume to half and raised to a pH of 2. The solution is brought to boiling and NaCl solution is added to precipitate the AgCl. After two hours, the AgCl is filtered, washed and dried for a minimum of 45 minutes at 130°C. Estimation of silver is made volumetrically after the method of FAJANS and WOLFE (1924). Titration is carried out against a 0.1 *N* KBr solution with rhodamine 6G (supplied by Chroma-Gesellschaft, Smith und Co., Stuttgart–Untertürkheim) as the absorption indicator.

Separations of silver from lead when the Ag: Pb ratio is as low as 1 : 10,000 are claimed. Fe^{3+} salts do not interfere so long as the Fe : Pb ratio does not exceed 100 : 1. Some results for silver are given in Table XXXIX.

Gold from transitional metals (ZIEGLER and MATSCHKE, 1962)

First, prepare the Ag_2S-cellulose by taking 25 g MN-cellulose-pulver 2,100 CdS-8% (of Macherey, Nagel und Co., Düren) and

TABLE XXXIX

SILVER ESTIMATIONS IN PREPARED SAMPLES

Silver		
Given (mg)	*Found (mg)*	*Error %*
43.7	43.5	−0.4
21.2	21.4	+0.9

saturating with water. Stir in an aqueous solution of $AgNO_3$ (8 g in 100 ml) and continue stirring for 30 minutes. Then decant or draw off by suction the supernatant liquid and wash several times with water. Wash with agitation to eliminate $AgNO_3$ and $Cd(NO_3)_2$ completely. The dark Ag_2S-cellulose is stored in polythene bottles under water, under which conditions it lasts 3 months. The quantity prepared is sufficient for thirty Au separations. The 2.5 cm column is packed as in the previous method, tube dimensions being 16 cm × 2 cm. Then proceed as follows.

The test material in the form of a nitric acid (HCl free) solution is adjusted to within the pH range 1.0–1.5 by adding NaOH solution or acid. Test solution is passed through the column at a rate of 200 ml per hour under the influence of suction applied to the lower end of the tube. After the passage of the sample solution the cellulose column is extruded mechanically or by air pressure. Ash either the whole of the cellulose or at least the dark part. Dissolve residue in HNO_3:HCl mixture. Evaporate to dryness to drive off the acid and take up the residue in a little concentrated HCl. Adjust pH to within the range 1–4 and extract gold with methylene chloride. Estimation of the gold is made "photometrically" on the non-aqueous phase using the colour developed with polyethyleneglycol gold bromide [polyäthylenglykoloxonium–bromoaurat (III)] measurements being made at 394 mμ (Spektraphotometer PMQII by Carl Zeiss was the instrument used).

The method is applicable to the separation of gold from large quantities of ferric compounds. No interference is encountered with Fe^{3+} even up to 30 g/litre. Separations of gold from lead (with a mean error of 1.5%) are possible from solutions containing a gold: lead ratio of 1:200,000. Some typical results for gold are given in Table XL.

TABLE XL

GOLD ESTIMATIONS IN PREPARED SAMPLES

Gold		
Given (μg)	*Found (μg)*	*Error %*
710	700	−1.5
284	283	−0.4
142	144	+1.4

CHROMATOGRAPHIC PROCESSES IN GEOLOGY

The roles of such chromatographic processes as diffusion, adsorption, and ion-exchange, in the important geological processes of mineral genesis, sedimentation and diagenesis have received considerable attention from time to time. Any modern approach to these problems should be based on an understanding of the interrelation of the chromatographic processes and the yet unsolved complexity of the simple chromatographic systems.

Early investigators (such as LIESEGANG, 1913; HATSCHEK, 1922; BOYDELL, 1924; ZSIGMONDY, 1917; LINDGREN, 1919) made significant contributions to the study of the role of gels in the formation of minerals. In general, they oversimplified what are now regarded as complex processes and they generalised in ways that one would hesitate to do now. Enthusiasm was succeeded by caution and the contributions of SCHOUTEN (1934), BASTIN (1950), SCHNEIDERHÖHN and RAMDOHR (1931) and EDWARDS (1954) were more realistic and less speculative.

The strict scrutiny of many so-called "amorphous colloids" (for

TABL

METAL ION M

Gel and pH	Molecular solution[1]			
	Cation and radius	Anion and radius	pH	Molarity
Silica 10.4±0.1	Co^{2+} 0.72	Cl^- 1.81	2.5	1M
	Ni^{2+} 0.69	Cl^-	5.4	1M
	Fe^{3+} 0.64	Cl^-	0.7	1M
	Sn^{2+} 0.93	Cl^-	0.8	1M
	Cu^{2+} 0.72	Cl^-		1M
	Cu^{2+} 0.72	Cl^-		$^1/_2M$
	Cu^{2+} 0.72	Cl^-		$^1/_3M$
Alumina 7.6 ± 0.1	Co^{2+} 0.72	Cl^-	2.5	1M
	Ni^{2+} 0.69	Cl^-	5.4	1M
	Fe^{3+} 0.64	Cl^-	0.7	1M
	Sn^{2+} 0.93	Cl^-	0.8	1M
	Cu^{2+} 0.72	Cl^-		1M

[1] In some solutions, hydrochloric acid was added to prevent hydrolysis, hence the l pH values.

example, clays and other hydrolyzates) under X-ray analysis, showed most of them to be crystalline. The concept of the subordinate role of amorphous gels in the mineralogical composition (RANKAMA and SAHAMA, 1949, p.202) spread some doubts on their role in mineral genesis. Thus, in the last twenty years, the role of gels in ore mineral genesis has received less attention than one might have expected.

Meanwhile, from sedimentation studies, the role of gels and colloids in diagenesis was being recognised as increasingly important, especially in biochemical environments (WALKER, 1962; KRAUSKOPF, 1956; SIEVER, 1957, 1962; WHITE and CORWIN, 1961; ALDERMANN and SKINNER, 1957). In the problems of solution, concentration and transport of ore-forming materials, colloidal solutions seemed to offer some advantages over molecular solutions (NEIDLE, 1917; MOREY and FOURNIER, 1960; LINDGREN, 1919, p.125).

In the field of petrogenesis, the role of what the author chooses to

KLI

RATION IN GELS

Metal migration in mm									
Time in hours						*Time in days*			
1	*2*	*4*	*6*	*8*	*18*	*1*	*2*	*5*	*11*
4.0	5.8	8.3	9.5	11.1	16.6	19.2	27.8	43.5	63.6
3.2	4.8	6.6	9.0	10.1		17.7	25.8	40.1	61.0
8.6	15.6		35.0	42.5	59.0	67.2	91.2		
7.3	11.8	15.0	19.4	25.0	32.3	38.3	56.0	86.0	
4.6	6.6	9.4	10.8	12.5	18.4	25.8	34.0		
2.6	4.2	7.4	10.0	12.2	19.2	26.0	34.0		
2.4	4.1	7.2	9.4	10.2	14.1	17.1	24.0	36.0	
5.0	8.5	9.2	11.1	10.8					
5.0	18.5	24.0	27.0	31.4					
6.8	11.4	15.8	21.2	28.6					
6.0	9.0	14.0	18.6	26.4	34.8				
10.5	31.0	52.0							

call chromatographic processes is of the utmost importance (RAMBERG, 1952).

Advances in analytical techniques in ore microscopy and X-ray and differential thermal analyses, a more realistic appreciation of the chromatographic processes and the growing popularity of syngenesis in theories of ore formation have renewed interest in the role of gels in ore mineral genesis (RITCHIE, 1962b). The concept of the crystalline structure of colloidal particles is still in the stage of development (RANKAMA and SAHAMA, 1949; KLEIN, 1961, 1962). This concept should not be confused with the crystallinity assumed by gels upon aging. Deductive attempts at assessing the role of gels in mineral genesis rest uneasily on the interpretation of textures, especially the colloform texture. SCHOUTEN (1934), EDWARDS (1954) and BASTIN (1950) have emphasized strongly the dangers in such practices. Recent interpretations of this kind include those of LAVERTY and GROSS (1955), SWANSON (1961), WEEKS (1955), HEWITT and FLEISCHER (1960) and ROY (1959). In the opinion of the present author, the question of the validity of linking colloform texture with mineral genesis from a gel should be kept open and should be the subject of much more rigid investigations.

The present author, in some preliminary research into the diffusion of metal ions in gels, has attempted to prepare gels of silica, alumina and ferric hydroxide under conditions as closely as possible comparable to those existing in nature (RITCHIE, 1962b). Gels of uniform composition were produced in glass tubes($^1/_4$″ bore and 4 ft. lengths). The tubes were then cut into 8″ lengths to provide sets of six uniform gel samples. The tubes were immersed separately in molecular solutions containing separately the common metals in various concentrations. The tubes of gel and the molecular solutions were enclosed in larger glass tubes ($^3/_4$″ bore and 10″ long). Coagulation of the gels (observable through the glass) indicated the rate of diffusion. Colorimetric analyses of one millilitre samples of the molecular solutions made from time to time, permitted the estimation of the mass of the metal ions being taken up by the gels. The experiments were conducted first at room temperature and later at temperatures up to 90° C. Much more data are needed before conclusions should be drawn. However, there are some indications that the pH of the gel and of the molecular solution, the molarity of the molecular solution, the ionic radius of the metal and the temperature, influence the rate of diffusion. Liesegang rings often appear after one or two days and this probably indicates a secondary migration of the metal ions by diffusion. Some preliminary results are given in Table XLI.

CONCLUSION

It will be appreciated that the applications of chromatography to geology have been numerous. In routine analyses, in the laboratory and in the field, they have become equally attractive, and in some cases more attractive, than the classical analyses. In theoretical geology, the chromatographic processes have become recognised as being of the greatest importance.

Appendix

There are included in this section brief comments on some papers which could not be included conveniently in the main text and on others of foreign origin or of most recent publication that have just become available to the author.

Jedwab, J., 1958. Adaption of paper chromatography to chemical prospecting for uranium deposits of the Katanga type. *Proc. U.N. Intern. Conf. Peaceful Uses At. Energy, 2nd, Geneva, 1958*, 2: 178–181.

This paper chromatographic method of geochemical prospecting for uranium, cobalt, nickel and copper is essentially a modification of the method of Hunt et al. (1955).

Almond, H., Stevens, R. E. and Lakin, H. W., 1953. A confined-spot method for the determination of traces of silver in soils and rocks. *U.S. Geol. Surv., Bull.*, 992 (7): 71–81.

After solution of the sample, silver is separated by a solvent extraction method. The dispersed silver is concentrated into a spot on a chromatogram. Estimation is made by comparison with standard chromatograms.

Bloom, H., 1953. Methods used in geochemical prospecting analytical laboratory. *Econ. Geol.*, 48: 322–323.

Chromatographic methods for Ni, Cu and Ag are mentioned.

Ostle, D., 1954. Geochemical prospecting for uranium. *Mining Mag., London*, 91: 201–208.

An anion exchange resin is used to concentrate uranium from natural waters. A fluorometric determination follows.

BROWN, J., GRANT, C. L., UZOLUNE, S. C. and LEDROW, J. C., 1962. Mineral composition of some drainage waters from arctic Alaska. *J. Geophys. Res.*, 67:2447–2453.

Water samples are collected on an ion-exchange resin column. Recovery values up to 96% are claimed for Na, K, Mg, Ca and Ba.

SHAPIRO, L., 1960. Referred to in: *U.S. Geol. Surv., Profess. Papers*, 400A: 71.

In a method for the estimation of fluorine in phosphate rock, the sample is taken up in dilute nitric acid and then passed through a cation exchange resin column. Fluorine in the effluent is determined by its bleaching action on the red aluminium-alizarin complex.

ALMOND, H. and BLOOM, H., 1951. A semi-micro method for the determination of cobalt in soils and rocks. *U.S. Geol. Surv., Circ.*, 125: 1–6.

An early paper chromatographic method.

NAGLE, R. A. and MURTHY, T. K. S., 1959. Ion-exchange method for the separation of thorium from rare-earths and its application to monazite analysis. *Analyst*, 84: 37–41.

Thorium from the sample solution is sorbed onto Amberlite IRA-400 $(SO_4)^{2-}$ from a sulphate solution at pH 2. Elution is with 2 *N* HCl. Determination is made gravimetrically from the oxalate.

STREET, K. and SEABORG, G. T., 1948. The ion-exchange separation of zirconium and hafnium. *J. Am. Chem. Soc.*, 70: 4268–4269.

Hafnium is separated from zirconium firstly by the sorption of both on Dowex-50 and later by elution of each separately by 6 *M* HCl. A slight overlap in the composition of the eluant fractions is reported.

POLLARD, F. H., MCOMIE, J. F. W. and MARTIN, J. V., 1956. The quantitative analysis of the alkaline-earth metals by paper

chromatography. *Analyst*, 81: 353–358.
Mixtures of microgram amounts of barium, strontium, calcium and magnesium are separated as formates. Descending development is employed using the solvent methanol (50 ml), isopropanol (30 ml), formic acid (2 ml), water (15 ml) and ammonium formate (2.5 g). After separation on the paper, the metals are extracted and determined spectrophotometrically. Successful analyses have been made on dolomites and limestones.

NISHIMURA, M. and SANDELL, E. B., 1961. Photometric determination of zinc in meteorites. *Anal. Chim. Acta*, 26: 242–248.
Samples from silicate meteorites are decomposed with HF and H_2SO_4 while those from iron meteorites are decomposed with HCl and HNO_3 (and a sodium carbonate fusion, if necessary). Final solutions in each case are made with 2 *N* HCl. The zinc is separated from such metals as iron, nickel, copper and cobalt by passing the acid solution through a 3 cm $\times$ 1 cm column of Dowex 1-X8 resin. Zinc is eluted with 0.001 *N* HCl, the other metals mentioned with 2 *N* HCl. Platinum and similar metals are retained on the resin. Zinc is determined spectrophotometrically.

RILEY, J. P., 1958. The rapid analysis of silicate rocks and minerals. *Anal. Chim. Acta*, 19: 413–428.
Sodium and potassium in silicate rocks are determined after the removal of iron, aluminium and titanium by means of Amberlite IRA-400 in the citrate form.

KALLMANN, S., OBERTHIN, H. and LIU, R., 1962. Determination of niobium and tantalum in minerals, ores and concentrates using ion-exchange. *Anal. Chem.*, 34: 609–613.
Solution of the sample is achieved either by HF–HCl attack or by $NaHSO_4$ fusion. After sorption on a Dowex-1 column, unwanted metals are eluted with $HCl:HF:H_2O = 5:4:11$. Niobium is eluted with NH_4Cl–HF mixture and determined by gravimetric or photometric methods. Tantalum is eluted with NH_4Cl–NH_4F and determined in the same way.

POVONDRA, P. and CECH, F., 1962. The rapid analysis of natural phosphates. *Chem. Anal., Warsaw,* 51:37–39.

An ion-exchange method for the separation of zinc, iron, aluminium, manganese, magnesium, calcium and other metals is described.

COULOMB, R., 1962. Dosage de quelques oligoéléments dans les matériaux géologiques par irradiation neutronique et chromatographie sur papier. *Compt. Rend.*, 254: 4328–4329.

Neutron radiation and paper chromatography are combined to provide two methods for the estimation of the metals Mn, Sc, Cu, Co, Ga, W, Ta, Th and U in rocks. For metals with medium to long half-life periods the specimens are irradiated and then chromatographed. For metals of short half-life period this order is reversed.

AGRINIER, H., 1962. Détermination semi-quantitative de l'or dans les minéraux, les sols et les roches par chromatographie ascendante sur papier. *Compt. Rend.*, 255: 2801–2803.

After the sample is taken into solution by a procedure involving aqua regia, the mixture is centrifuged and aliquots up to 0.5 ml are chromatographed with the solvent mixture ethanol (20), ethyl acetate (20), water (20) and nitric acid (s.g. 1.33) (0.7). The detecting reagent is a saturated solution of p-dimethylaminobenzylidene-rhodanine in a 1:1 mixture of ethanol and acetone. A different method of sample solution is recommended for lateritic soils.

SULCEK, Z., DOLEZAL, J. and MICHAL, J., 1961. Analytische Schnellmethoden zur Untersuchung von Metallen und anorganischen Rohstoffen. 12. Bestimmung der Spurenmengen von Beryllium in Mineralwässern und mineralischen Rohstoffen. *Collection Czech. Chem. Commun.*, 26: 246–254.

Trace amounts of beryllium in rocks and natural waters are determined fluorometrically after selective separation on a silica gel column.

POVONDRA, P. and SULCEK, Z., 1959. Analytische Schnellmethoden zur Untersuchung von Metallen und anorganischen Rohstoffen 10. Anwendung der Ionenaustauscher zur Bestimmung von Mangan und alkalischen Erden. *Collection Czech. Chem. Commun.*, 24: 2398–2404.

Manganese and the alkaline earths are sorbed on Amberlite 1R-120. Elution is effected with E.D.T.A. under different pH conditions.

SULCEK, Z., MICHAL, J. and DOLEZAL, J., 1959a. Schnellmethoden in der Analyse von Metallen und mineralischen Rohstoffen. 8. Bestimmung kleiner Uranmengen in mineralischen Rohstoffen. *Collection Czech. Chem. Commun.*, 24: 1815–1821.

SULCEK, Z., MICHAL, J. and DOLEZAL, J., 1961b. Sensitive method for detection and semiquantitative determination of uranium. *Chem. Anal., Warsaw*, 50: 13–14.

A rapid method for the separation of uranium from associated metals is achieved by complexing with E.D.T.A. and tartrate on a silica gel column.

SULCEK, Z., POVONDRA, P. and STANGL, R., 1962. Untersuchung der Kationensorption aus Komplexanmedium-III. Chromatographische Trennung von Strontium und Barium. *Talanta*, 9: 647–651.

A new method is presented for the separation of Sr, Ca and Ba by sorption on a column of Amberlite 1R-120 in NH_4^+ form. Elution is effected with 0.02–0.05 *M* DCyTE (1,2-diaminocyclohexane-N-N, N′, N′-tetra-acetic acid) in 0.4 *M* ammonium acetate in the pH range 6.3–7.2 for Sr and at pH 5.1 for Ca. The method is applicable to the estimation of Sr and Ca in barium minerals such as barite and witherite.

POVONDRA, P. and SULCEK, Z., 1961. Determination of calcium in strontium salts and strontium in calcium-rich materials. *Chem. Anal., Warsaw*, 50: 79 and 93–94.

Separation is achieved by four methods (according to the nature of the separation and the sample) involving sorption on and later elution from a strongly acidic cation exchange resin. One method is suitable for the determination of trace amounts of strontium in calcite, aragonite and mineral waters.

KORKISCH, J. and JANAUER, G. E., 1962. Ion exchange in mixed solvents. Separation methods for uranium and thorium. *Talanta*, 9: 957–985.

The anion-exchange behaviour of U^{6+} and Th^{4+} in mixed solvents is discussed. Seven separation methods are presented. Two of the methods are directly applicable to the determination of thorium and one of these two for the determination of uranium. Dowex resin columns are recommended.

KORKISCH, J., FARAG, A. and HECHT, F., 1958. Eine neue Methode zur Anreicherung des Urans mittels Ionenaustausches und deren Anwendung zur Bestimmung des Urans in festen Proben. *Mikrochim. Acta*, 3: 415–425.

An ion-exchange method for the separation of uranium as a negatively-charged ascorbate complex from a large number of associated metals on an Amberlite IRA-400 (ascorbate) resin.

KORKISCH, J., THIARD, A. and HECHT, F., 1956. Schnellbestimmung des Urans in Meer- und Flusswässern. *Mikrochim. Acta*, 9: 1422–1430.

Ion-exchange column methods for sampling sea water and river water are described. Amberlite IRA-400 (in the Cl^- or Ac^- forms respectively) is used. Uranium is eluted with 0.8 N HCl and determined polarographically.

HECHT, F., KORKISCH, J., PATZAK, R. and THIARD, A., 1956. Zur Bestimmung kleinster Uranmengen in Gesteinen und natürlichen Wässern. *Mikrochim. Acta*, 7–8: 1283–1309.

The separation of uranium from other elements is effected after an ether extraction from a nitric acid solution on Amberlite 1RA-400 (Ac^-). Elution is achieved with 0.8 N HCl. The uranium is determined polarographically.

KORKISCH, J., ANTAL, P. and HECHT, F., 1959. Bestimmung von Uran und Thorium in naturlichen Wässern nach vorangehender Anreicherung an Amberlite 1RA-400 and Dowex-50. *Mikrochim. Acta*, 5: 693–705.

A method of sampling and estimating the uranium and thorium content of natural waters is described. The two elements with other metals are sorbed onto Amberlite 1RA-400 (ascorbate form) and later separated from interfering metals by combining the resin already mentioned with Dowex-50 (H^+ form) and selective eluants. Final determinations are made polarographically.

FERGUSON, W. S., 1962. Analytical problems in determining hydrocarbons in sediments. *Bull. Am. Assoc. Petrol. Geologists*, 46: 1613–1620.

After hydrocarbons are extracted from the sediments by benzene they are separated on silica gel columns.

SEIM, H. J., MORRIS, R. J. and FREW, D. W., 1957. Rapid routine method for the determination of uranium in ores. *Anal. Chem.*, 29: 443–446.

After a mineral acid dissolution, the uranium and associated metals are sorbed onto anion exchange resins (Amberlite 1RA-400 or Dowex-2). Elution is achieved with hot 1 *M* perchloric acid. Uranium is determined colorimetrically.

ELBEIH, I. I. M. and ABOU–ELNAGA, M. A., 1958a. A new method based on paper chromatography for the estimation of thorium and uranium in monazite. *Anal. Chim. Acta*, 19: 123–128.

Thorium and uranium are separated from the sample solution and each other by paper chromatography. Uranium is estimated spectrophotometrically and thorium indirectly.

ELBEIH, I. I. M. and ABOU–ELNAGA, M. A., 1958b. A new method based on paper chromatography for the determination of uranium in uranium minerals. *Chem. Anal., Warsawa*, 47: 92–93.

After separation from associated metals by paper chromatography, the uranium spots are cut from the paper and uranium is determined by titration with E.D.T.A.

UMEZAKI, Y., 1958. Determination of zinc in iron ores in the presence of cobalt. *Bunseki Kagaku*, 7: 37–42. Abstract in: *Anal. Abstr.*, 1958: 3636.

Ion exchange with Dowex 1-X8 is used to separate zinc from iron and cobalt. A polarographic determination of zinc follows.

References

AGRINIER, H., 1957a. Applications de la chromatographie ascendante de partage sur papier à la détermination de certains éléments dans les minéraux. 1. Détermination du lithium, du bore et du béryllium. *Bull. Soc. Franç. Minéral. Crist.*, 80:181–193.

AGRINIER, H., 1957b. Applications de la chromatographie ascendante de partage sur papier, à la détermination de certains éléments dans les minéraux. 2. Détermination et estimation semi-quantitative de l'argent, du nickel, du cobalt, du cuivre, du niobium, du tantale et du titane. *Bull. Soc. Franç. Minéral. Crist.*, 80:275–292.

AGRINIER, H., 1959. Chromatographie sur papier du molybdène. Application à la détermination semi-quantitative de cet élément dans les minéraux, les roches et les sols. *Compt. Rend.*, 249:2365–2366.

AGRINIER, H., 1960. Dosage semi-quantitatif du béryllium dans les roches et les sols. *Chim. Anal. Paris*, 12:600–602.

AGRINIER, H., 1961a. Détermination semi-quantitative de l'arsenic dans les minéraux par chromatographie ascendante sur papier. *Compt. Rend.*, 253: 1980–1981.

AGRINIER, H., 1961b. Détermination semi-quantitative rapide du bismuth dans les minéraux par chromatographie ascendante sur papier. *Compt. Rend.*, 253:280–281.

AGRINIER, H., 1962. Détermination semi-quantitative du sélénium dans les minéraux et les sols par chromatographie ascendante sur papier. *Compt. Rend.*, 254:1850–1851.

ALBERTI, G. and GRASSINI, G., 1960. Chromatography on paper impregnated with zirconium phosphate. *J. Chromatog.*, 4:83–85.

ALDERMANN, A. R. and SKINNER, H. C. W., 1957. Dolomite sedimentation in the southeast of South Australia. *Am. J. Sci.*, 255:561–567.

ALLEN, J. A. and PICKERING, W. F., 1961. The sorption of bromine by polythene. *Australian J. Appl. Sci.*, 12:42–50.

AMBROSE, D. and AMBROSE, B. A., 1961. *Gas Chromatography*. George Newnes, London, 220 pp.

BASOLO, F., LEDERER, M., OSSICINI, L. and STEPHEN, K. H., 1963. A paper chromatographic study of some platinum (11) compounds. 2. The separation of cis- and trans-dihalogenodiamminoplatinum (11) complexes. *J. Chromatog.*, in press.

BASTIN, E. S. 1950. Interpretation of ore textures. *Geol. Soc. Am.,Mem.*,45: 101 pp.

BAYER, E., 1961. *Gas Chromatography.* Elsevier, Amsterdam, 240 pp.

BELCHER, C. B., 1963. Sodium peroxide as a flux in refractory and mineral analysis. *Talanta,* in press.

BLOCK, R. J., DURRUM, E. L. and ZWEIG, G., 1958. *A Manual of Paper Chromatography and Electrophoresis.* Academic Press, New York, 710 pp.

BOYDELL, H. C., 1924. The role of colloidal solutions in the formation of mineral deposits. *Trans. Inst. Mining Met.,* 34:145–337.

BRIMLEY, R. C. and BARRETT, F. C., 1954. *Practical Chromatography.* Chapman and Hall, London, 128 pp.

BRITISH DRUG HOUSES, 1958. *B.D.H. Book of Organic Reagents.* British Drug Houses Ltd., Poole, 175 pp.

BUERGER, M. J., 1960. *Crystal Structure Analysis.* Wiley, New York, 668 pp.

CANNEY, F. C. and HAWKINS, D. B., 1960. Field application of ion-exchange resins in hydrogeochemical prospecting. *U.S. Geol. Surv., Profess. Papers,* 400-B: 89–90.

CARRITT, D. E., 1953. Separation and concentration of trace metals from natural waters. *Anal. Chem.,* 25:1927–1928.

CASSIDY, H. G., 1951. *Adsorption and Chromatography.* Interscience, New York, 370 pp.

CETINI, G. and RICCA, F., 1955–1956. Caratteristiche cromatografiche del silicagel basico. *Atti Accad. Sci. Torino, Classe Sci. Fis., Mat. Nat.,* 90:229–242.

CONSDEN, R., GORDON, A. H. and MARTIN, A. J. P., 1944. Qualitative analysis of proteins: a partition chromatographic method using paper. *Biochem. J.,* 38:224–232.

COULOMB, R., 1957. Prospection géochimique en Guyane. *Comm. Énergie At., France, Note,* 227: 3–10.

COULOMB, R. and GOLDSTEIN, M., 1956a. Recueil de méthodes de dosage de l'uranium utilisées en géochimie. *Comm. Énergie At., France, Note,* 156:1–17.

COULOMB, R. and GOLDSTEIN, M., 1956b. Prospection et recherche de l'uranium. Les techniques annexes. Prospection géochimique. *Rev. Ind. Minérale,* 1-R: 140–153.

DAL NOGARE, S. and JUVET, R. S., 1962. *Gas-Liquid Chromatography.* Interscience, New York, 450 pp.

DANA, E. S., 1932. *A Textbook of Mineralogy,* 4th ed. Wiley, New York, 851 pp.

DANIELS, F., 1951. *Outlines of Physical Chemistry.* Wiley, New York, 713 pp.

DOBSON, E. M., 1962. Analyses of ruby and sapphire maser crystals. *Anal. Chem.,* 34:966–971.

DOW CHEMICAL COMPANY, year of publication unknown. *Dowex Ion Exchange.* Dow Chemical Company, Midland, 80 pp.

DYKYJ, J. and CERNY, J., 1945. A new method of qualitative chromatography of inorganic cations. *Chem. Listy,* 39:84–91.

EDWARDS, A. B., 1954. *Textures of the Ore Minerals.* Australasian Inst. Mining and Met., Melbourne, 242 pp.

ELLINGTON, F. and ADAMS, W. N., 1951. Determination of phosphorus in coal. *Fuel,* 30:272–274.

ELLINGTON, F. and STANLEY, N., 1955. The use of ion-exchange resins in the analysis of coal ash. *Analyst*, 80:313–315.

FAJANS, K. and WOLFE, H., 1924. Titration of Ag and halogen ions with organic dyestuff indicators. *Z. Anorg. Allgem. Chem.*, 137:221–245.

FEIGL, F., 1954. *Spot Tests*. Elsevier, Amsterdam, 518 pp.

FINDLAY, A., 1942. *Introduction to Physical Chemistry*. Longmans–Green, London, 582 pp.

FLASCHKA, H. A., 1959. *E.D.T.A. Titrations*. Pergamon, London, 138 pp.

GLASSTONE, S. and LEWIS, D., 1960. *Elements of Physical Chemistry*. Van Nostrand, Princeton, 758 pp.

GORDON, A. H., MARTIN, A. J. P. and SYNGE, R. L. M., 1943. Partition chromatography in the study of protein constituents. *Biochem. J.*, 37:79–86.

GREEN, J., 1959. Geochemical table of the elements for 1959. *Bull. Geol. Soc. Am.*, 70:1127–1184.

HALE, D. K., 1955. A new technique in ion-exchange chromatography. *Chem. Ind. London*, 1955:1147–1148.

HANES, C. S. and ISHERWOOD, F. A., 1940. Separation of the phosphoric esters on the filter paper chromatogram. *Nature*, 164:1107–1112.

HATSCHEK, E., 1922. *Introduction to the Physics and Chemistry of Colloids*. Churchill, London, 172 pp.

HEWITT, D. F. and FLEISCHER, M., 1960. Deposits of manganese oxides. *Econ. Geol.*, 55:1–55.

HJELLE, A., 1961. Forsøk på kromatografisk bestemmelse av (Ca–Mg)-karbonater. *Norg. Geol. Undersøkelse*, 213:58–61.

HODGMAN, C. D., 1958. *Handbook of Chemistry and Physics*. Chemical Rubber Co., Cleveland, 3456 pp.

HUNT, E. C. and WELLS, R. A., 1954. Inorganic chromatography on cellulose. 15. A rapid chromatographic method for the determination of niobium in low-grade samples. *Analyst*, 79:351–359.

HUNT, E. C., NORTH, A. A. and WELLS, R. A., 1955. Application of paper-chromatographic methods of analysis to geochemical prospecting. *Analyst*, 80:172–194.

JERMYN, M. A. and ISHERWOOD, F. A., 1949. Improved separation of sugars on a paper partition chromatogram. *Biochem. J.*, 44:402–407.

JOST, W., 1960. *Diffusion*. Academic Press, New York, 558 pp. (plus appendix 94 pp.).

KEMBER, N. F. and WELLS, K. A., 1951. Inorganic chromatography on cellulose. 6. Extraction and determination of gold. *Analyst*, 76:579–587.

KETELAAR, J. A. A., 1958. *Chemical Constitution*. Elsevier, Amsterdam, 448 pp.

KIRCHNER, J. G., MILLER, J. M. and KELLER, G. J., 1951. Separation and identification of some terpenes by a new chromatographic technique. *Anal. Chem.*, 23:420–425.

KLEIN, P. D., 1961. Silica gel structure and the chromatographic process. *Anal. Chem.*, 33:1737–1741.

KLEIN, P. D., 1962. Silica gel structure and the chromatographic process — surface energy and activation procedures. *Anal. Chem.*, 34:733–736.

KNIGHT, C. S., 1959. A two-dimensional paper chromatographic method combining ion-exchange and partition techniques. *Nature*, 183:165–167.

KRAUS, K. A. and MOORE, G. E., 1950. Adsorption of iron by anion exchange resins from hydrochloric acid solutions. *J. Am. Chem. Soc.*, 72:5792–5793.

KRAUSKOPF, K. B., 1956. Dissolution and precipitation of silica at low temperatures. *Geochim. Cosmochim. Acta*, 10:1–26.

KRIEGER, K. A., 1941. Adsorption. 1. The effect of heat-treatment on the low-temperature adsorption of nitrogen by aluminium oxide. *J. Am. Chem. Soc.*, 63:2712–2714.

KUNIN, R., 1958. *Ion Exchange Resins*. Wiley, New York, 466 pp.

LACOURT, A., SOMMEREYNS, G., DEGEYNDT, E., BARUH, J. and GILLARD, J., 1949. Separatory power of organic solvents in quantitative chromatographic separations of inorganic compounds present in μ quantities. *Mikrochim. Acta*, 34:215–233.

LAVERTY, R. A. and GROSS, E. B., 1955. Paragenetic studies of uranium deposits of the Colorado Plateau. *U.S. Geol. Surv., Profess. Papers*, 300:195–201.

LEDERER, M., 1953. Separation of the rare earths by paper chromatography. *Compt. Rend.*, 236:1557–1559.

LEDERER, M., 1955. Chromatography on paper impregnated with ion-exchange resins. Preliminary report. *Anal. Chim. Acta*, 12:142–145.

LEDERER, M. and KERTES, S., 1956. Chromatography on paper impregnated with ion-exchange resins. 2. Separation of selenite and tellurite. *Anal. Chim. Acta*, 15:226–231.

LEDERER, E. and LEDERER, M., 1957. *Chromatography*. Elsevier, Amsterdam, 711 pp.

LIESEGANG, R. E., 1913. *Geologische Diffusionen*. Steinkopff, Dresden, 180 pp.

LINDGREN, W., 1919. *Mineral Deposits*, 2nd ed. McGraw–Hill, New York, 957 pp.

LITTLEWOOD. A. B., 1962. *Gas Chromatography*. Academic Press, New York, 507 pp.

MILTON, R. F. and WATERS, W. A., 1955. *Methods of Quantitative Microanalysis*. Arnold, London, 742 pp.

MISTRYUKOV, A. E., 1962. Thin layer chromatography using the descending technique with non-bound alumina plates. *J. Chromatog.*, 9:311–313.

MOREY, G. W. and FOURNIER, R. O., 1960. As reported in *U.S. Geol. Surv., Profess. Papers*, 400-A: 61.

NEIDLE, M., 1917. The precipitation, stability and constitution of hydrous ferric oxide sols. *J. Am. Chem. Soc.*, 39:2334–2350.

NEVILL, H. F. C. and LEVER, R. R., 1959. Prospecting for fine gold using a paper chromatographic method. *Australasian Inst. Mining Met., Proc.*, 191:141–164.

NORTH, A. A. and WELLS, R. A., 1959. Analytical methods for geochemical prospecting. *Intern. Geol. Congr., 20th, Mexico, 1956, Rept., Symp. Exploracion Geoquimica*, 2:347–362.

PARTINGTON, J. R., 1949. *An Advanced Treatise on Physical Chemistry*. Longmans, Green, London, 1:943 pp.

PERMUTIT Co., year of publication unknown. *"Permutit" Ion Exchange Resins in the Laboratory*. The Permutit Co. Ltd., London, 25 pp.

PETSCHIK, H. and STEGER, E., 1962. Chromatographische Trennung einfacher aliphatischer Thiophosphorsaureester mittels Dunnschichtchromatographie. *J. Chromatog.*, 9:307–310.

PICKERING, W. F., 1959. *The Role of Adsorption and Complex Formation in Inorganic Paper Chromatography*. Ph. D. thesis, Univ. of New South Wales, Newcastle (unpublished).

POLLARD, F. H. and MCOMIE, J. F. W., 1953. *Chromatographic Methods of Inorganic Analysis*. Butterworths, London, 192 pp.

RAFTER, T. A., 1950. Sodium peroxide decomposition of minerals in platinum vessels. *Analyst*, 75:485–492.

RAMBERG, H., 1952. *The Origin of Metamorphic and Metasomatic Rocks*. Univ. Chicago Press, Chicago, 317 pp.

RANKAMA, K. and SAHAMA, T. G., 1949. *Geochemistry*. Univ. Chicago Press, Chicago, 911 pp.

REIO, L., 1958. A method for the paper-chromatographic separation and identification of phenol derivatives, mould metabolites and related compounds of interest, using a "reference system". *J. Chromatog.*, 1:338–373.

RITCHIE, A. S., 1961. A paper chromatographic scheme for the identification of metal ions. *J. Chem. Educ.*, 38:400–405.

RITCHIE, A. S., 1962a. The identification of metal ions in ore minerals by paper chromatography. 1. Opaque ore minerals. *Econ. Geol.*, 57:238–247.

RITCHIE, A. S., 1962b. *The Role of Gels in Ore Mineral Genesis*. Address to Australian New Zealand Assoc. Advan. Sci., Sect. C, Symp. 5 (unpublished).

ROCKLAND, L. B., BLATT, J. L. and DUNCAN, M. S., 1951. Small-scale filter-paper chromatography. *Anal. Chem.*, 23:1142–1146.

ROY, S., 1959. Mineralogy and texture of manganese bodies of Dongari Buzary, Bhandora District, Bombay State, India, *Econ. Geol.*, 54:1556–1574.

SALMON, J. E. and HALE, D. K., 1959. *Ion Exchange*. Butterworths, London, 136 pp.

SANDELL, E. B., 1944. *Colorimetric Determination of Traces of Metals*. Interscience, New York, 500 pp.

SCHAY, G., 1960. *Theoretische Grundlagen der Gaschromatographie*. VEB Deutscher Verlag der Wissenschaften, Berlin, 267 pp.

SCHNEIDERHÖHN, H. and RAMDOHR, P., 1931. *Lehrbuch der Erzmikroskopie*. Bornträger, Berlin, 714 pp.

SCHOUTEN, C., 1934. Structures and textures of synthetic replacements in "open space". *Econ. Geol.*, 29:611–658.

SHAPIRO, L. and BRANNOCK, W. W., 1956. Rapid analysis of silicate rocks. *U.S. Geol. Surv. Bull.*, 1036-C: 19–56.

SHEEHAN, W. F., 1961. *Physical Chemistry*. Allyn and Bacon, Boston, 618 pp.

SIEVER, R., 1957. The silica budget in the sedimentary cycle. *Am. Mineralogist*, 42:821–841.

SIEVER, R., 1962. Silica solubility, 0–200° C and diagenesis of sediments. *J. Geol.*, 70:127–150.

SMALES, A. A. and WAGER, L. R., 1960. *Methods in Geochemistry*. Interscience, London, 464 pp.

SMITH, O. C., 1953. *Inorganic Chromatography*. Van Nostrand, New York, 134 pp.

SPAIN, J. D., LUDEMAN, F. L. and SNELGROVE, A. K., 1962. The use of precipitation chromatography in geochemical prospecting; mineral identification with disposable agar gel columns. *Econ. Geol.*, 57:248–259.

STAHL, E., SCHRÖTER, G., KRAFT, G. and RENZ, R., 1956. Thin layer chromatography (the method, affecting factors, and a few examples of application). *Pharmazie*, 11:633–637.

STEVENS, H. M., 1959. The effect of the electronic structure of the cation upon fluorescence in metal-8-hydroxyquinoline complexes. *Anal. Chim. Acta*, 20: 389–396.

STRAIN, H. H., 1942. *Chromatographic Adsorption Analysis*. Interscience, New York, 222 pp.

SWANSON, V. E., 1961. Geology and geochemistry of uranium in marine black shales. *U.S. Geol. Surv., Profess. Papers*, 356-C: 1–112.

THOMPSON, C. E. and LAKIN, H. W., 1957. A field chromatographic method for determination of uranium in soils and rocks. *U.S. Geol. Surv. Bull.*, 1036-L: 209–220.

TURNER, F. J. and VERHOOGEN, J., 1951. *Igneous and Metamorphic Petrology*. McGraw–Hill, New York, 602 pp.

VAN ARKEL, A. E., 1956. *Molecules and Crystals*. Butterworths, London, 270 pp.

VAN ERKELENS, F. C., 1961. On the concentration and separation of the trace-elements Fe, Cu, Zn, Mn, Pb, Mo and Co. 3. Paper chromatography. *Anal. Chim. Acta*, 25:226–232.

WALDI, D., year of publication unknown. *Chromatography*. Merck, Darmstadt, 185 pp.

WALKER, T. R., 1962. The reversible nature of chert-carbonate replacement in sedimentary rocks. *Bull. Geol. Soc. Am.*, 73:237–242.

WARD, F. N. and MARRANZINO, A. P., 1957. Field determination of uranium in natural waters. *U.S. Geol. Surv. Bull.*, 1036-J: 181–192.

WARD, F. N., NAKAGAWA, H. M. and HUNT, B., 1960. Geochemical investigations of molybdenum at Nevares Spring in Death Valley, California. *U.S. Geol. Surv., Profess. Papers*, 400-B (207): 454–456.

WEATHERLEY, E. G., 1956. Chromatographic separation, detection and determination of selenium. *Analyst*, 81:404–408.

WEEKS, A. B., 1955. Mineralogy and oxidation of the Colorado Plateau uranium ores. *U.S. Geol. Surv., Profess. Papers*, 300:187–193.

WELCHER, F. J., 1947. *Organic Analytical Reagents*. Van Nostrand, Princeton, 1–4:2189 pp.

WELCHER, F. J. and HAHN, R. B., 1955. *Semimicro Qualitative Analysis*. Van Nostrand, Princeton, 497 pp.

WHITE, J. F. and CORWIN, J. F., 1961. Synthesis and origin of chalcedony. *Am. Mineralogist*, 46:112–119.

ZECHMEISTER, L. and CHOLNOKY, L., 1950. *Principles and Practice of Chromatography*. Chapman and Hall, London, 361 pp.

ZIEGLER, M., 1961. Die Fällung der Übergangsmetall-Ionen mit Metall- und Halbmetallsulfiden. Die Trennung des Kupfers von Blei mit Cadmiumsulfid. *Z. Anal. Chem.*, 180:4–8.

ZIEGLER, M., and GIESELER, M., 1962. Die Fällung von Übergangsmetallionen mit Metall- und Halbmetallsulfiden. 2. Die Trennung des Silbers von Blei durch Fällung mit Cadmiumsulfid. *Z. Anal. Chem.*, 191:122–126.

ZIEGLER, M. and MATSCHKE, H., 1962. Die Fällung von Übergangsmetallen mit Metall- und Halbmetallsulfiden. 3. Die Trennung des Goldes von Übergangsmetallen durch Fällung mit Silbersulfid. *Z. Anal. Chem.*, 191:188–190.

ZSIGMONDY, R., 1917. *Chemistry of Colloids*. Wiley, New York, 274 pp.

Index

PRINTED IN THE NETHERLANDS
BY N.V. BOEKDRUKKERIJ F.E. MAC DONALD, NIJMEGEN